Circadian Rhythm - New Insights Into Physiological and Pathological Implications

Edited by Cristina Manuela Drăgoi, Alina Crenguța Nicolae and Ion-Bogdan Dumitrescu

Published in London, United Kingdom

Circadian Rhythm - New Insights Into Physiological and Pathological Implications
http://dx.doi.org/10.5772/intechopen.96835
Edited by Cristina Manuela Drăgoi, Alina Crenguţa Nicolae and Ion-Bogdan Dumitrescu

Contributors
Michael Jackson Oliveira de Andrade, Elena A. Gromova, Valery V. Gafarov, Igor V. Gagulin, Vladimir N. Maksimov, Almira V. Gafarova, Gerard A. Kennedy, Robyn Hudson, Gloria Benítez-King, Rosa Estrada-Reyes, Jairo Muñoz-Delgado, Héctor Solís-Chagoyán, Salvador Alarcón-Elizalde, Cristina Manuela Drăgoi, Alina Crenguţa Nicolae, Ion Bogdan Dumitrescu, Camelia Cristina Diaconu, Mirela Elena Ritivoiu, Carmen Adella Sirbu

First published in London, United Kingdom, 2022 by IntechOpen
IntechOpen is the global imprint of INTECHOPEN LIMITED, registered in England and Wales, registration number: 11086078, 5 Princes Gate Court, London, SW7 2QJ, United Kingdom

British Library Cataloguing-in-Publication Data
A catalogue record for this book is available from the British Library

Additional hard and PDF copies can be obtained from orders@intechopen.com

Circadian Rhythm - New Insights Into Physiological and Pathological Implications
Edited by Cristina Manuela Drăgoi, Alina Crenguţa Nicolae and Ion-Bogdan Dumitrescu
p. cm.
Print ISBN 978-1-80355-369-6
Online ISBN 978-1-80355-370-2
eBook (PDF) ISBN 978-1-80355-371-9

Meet the editors

Cristina Manuela Drăgoi, MPharm, Ph.D., is a pharmacist and associate professor at the Department of Biochemistry, Faculty of Pharmacy, "Carol Davila" University of Medicine and Pharmacy, Romania. She has professional experience in oxidative stress mechanisms, assessing drugs' effects on biochemical markers, and identifying predictive biomarkers for metabolic diseases. Her scientific research focuses on revealing the circadian entrainment actions on major physiological and pathological processes and determining its modulatory abilities on different aspects of hormone synthesis, healthy ageing mechanisms, and the administration of nutrition and medicines. She has written several books, book chapters, and research articles in this area.

Alina Crenguța Nicolae, MPharm, Ph.D., is an associate professor at the Faculty of Pharmacy, "Carol Davila" University of Medicine and Pharmacy, Romania. She has conducted extensive research on individual neuro-biochemical profiles, with a focus on the predictive biomarkers generating the response to pharmacotherapy and the modulation of multidrug-resistance mechanisms. She has also conducted research on some biochemical and cardiovascular parameters in microbiology, chronobiology (chronopharmacology and chrononutrition), and circadian rhythms. She has published several research articles, books, and book chapters on biochemical aspects, enabling a better understanding of the process of multidrug resistance in living organisms and the circadian governance of cellular processes.

Ion-Bogdan Dumitrescu, MPharm, Ph.D., joined the Department of Pharmaceutical Physics and Informatics, "Carol Davila" University of Medicine and Pharmacy, Romania in 2007, where he is currently an assistant professor. Previously, he was the managing editor and webmaster for *Farmacia*, the official journal of the Romanian Society for Pharmaceutical Sciences. His doctoral thesis involved research on bio-relevant in-vitro release conditions for low-soluble pharmaceutical compounds. His main research interest is 3D printing for pharmaceutical preparations.

Contents

Preface

The evolution of life on Earth has a unique anticipatory quiddity that is intended to prepare and coordinate most living organisms to the daily environmental changes driven by the rotation of the planet. The constant transition from light to dark imposes sleep-wake patterns, but in addition to that, there are fluctuations in nutrient availability that impose supplementary pressure on the circadian cycle.

Most of the cells in complex organisms contain molecular clocks that control the activity of important signalling pathways. The synchronisation of these clocks to the circadian environment assures a physiologic pattern of the structure and functioning of all cells and tissues. Many physiological processes within the human organism are influenced by this circadian entrainment system in which the suprachiasmatic nucleus located in the brain acts as a master pacemaker that synchronises subsidiary clocks in peripheral cells. These biological clocks are assembled around gene expression, release modulation of signalling molecular triggers, and interlocking endocrine regulatory patterns, all in perfect coordination with rhythmical physiological and behavioural outputs.

Starting from an initial emphasis on sleep-wake cycles, feeding behaviour, and metabolism, continuing with the confirmation that circadian oscillations play a critical role in foremost physiological mechanisms, we are now witnessing a revolutionary step in chronobiology, namely, the establishment of new research realms that consider all exploratory approaches for the optimal pharmacological management of circadian oscillators and apply circadian principles to drug delivery strategies for therapeutic benefits. The book addresses these topics from the perspective of the implications of circadian rhythm in determining daily human activities, the chronobiologic echoes felt at the metabolic and endocrine levels, and in modulating pathologies outbursts and rationalising pharmacotherapy principles in a coherent individualised circadian pattern. The subject is of overwhelming importance, both for its unique applicability in everyday human life as well as for its extraordinary physio-pathological premises of being widely used to modulate environmental, nutritional, and pharmacological factors to improve quality of life and therapeutic results in perfect accordance with the unique biological rhythm of each individual. This book is a useful resource for healthcare professionals both clinically and academically interested in and impacted by research on circadian rhythms.

Cristina Manuela Drăgoi, Alina Crenguța Nicolae and Ion-Bogdan Dumitrescu
Faculty of Pharmacy,
Department of Biochemistry,
"Carol Davila" University of Medicine and Pharmacy,
Bucharest, Romania

Chapter 1

Introductory Chapter: Untangling the Essential Links among the Circadian Rhythm, Homeostasis of the Human Body, and the Nutritional, Behavioural, and Pathological Interferences

Cristina Manuela Drăgoi, Ion-Bogdan Dumitrescu and Alina Crenguța Nicolae

1. Introduction

The vast majority of living entities are subject to unavoidable and predictable conditions of 24-hour changes in their environment, having the ability to adjust to the day-night cycle. To manage these daily changes in periods of light and darkness, almost every living organism has developed an internal system of synchronisation or circadian clock. This tremendous discovery dates back to the beginning of the eighteenth century when Jean-Jacques d'Ortous de Mairan performed the first chronobiology experiment on *Mimosa pudica,* a plant that visibly responds to circadian changes [1, 2].

2. Circadian, infradian, and ultradian rhythms

Each living organism carries out its activity for approximately 24 hours by:

- obtaining energy (nutrients);
- optimising energy consumption by using it for daily activities and storing the rest for future needs;
- protection from predators;
- regeneration or growth processes;
- reproduction.

All these functions are channelled by a circadian rhythm, which adjusts the body's ability to perform the demanded tasks by assigning each role at an appropriate time of day or night. Even isolated from solar triggers, the biological clock keeps unaltered its endogenous period of about 24 hours under constant conditions. In actual life, circadian clocks are permanently entrained by external signals, the most significant being light, which also directs the daily rhythmicity of all other environmental cues.

Furthermore, bearing the utmost importance for humans, biological timing is essential for coordinating homeostasis, physiological processes, from gene expression to drug metabolism, and behaviour, in the same way as the disruption of biological synchronising mechanisms are evoked in significant pathological surges. The temporal information is mainly driven by the Earth's rotational cycle of 24 hours, imposing an equivalent rhythmic profile in almost every living entity, as profoundly engaged as the molecular level should be considered [1, 2].

In addition to circadian rhythms, the periodicity of certain events is described in more extended time frames, called generically infradian rhythms, which can last from several days, as is the case of the menstrual cycle, to annual cycles mainly found in animals that display infradian patterns in reproduction, hibernation or seasonal migration. Contrary-wise, the biological paces with a period shorter than 24 hours, starting from seconds or minutes to a few hours, are called ultradian rhythms. One of the most accurately described and studied ultradian rhythms is the 40 minutes cycle of cellular respiration displayed by *Saccharomyces cerevisiae* in aerobic culture conditions. Besides that, an interesting and early revealed human ultradian rhythm is the REM-NREM sleep sequence with an extent of 90 minutes and an occurrence of 3–5 cycles during normal sleep, also reflected on the recurrent functional pattern of muscles tone, brain electrical and oculomotor activity and nevertheless of energy utilisation. These rhythms are ubiquitously expressed in all biological systems and established in all organisms, from unicellular ones to mammals, from single cells to complex biological mechanisms.

3. The three time-setting entities: circadian endogenous clock—the solar clock—the social clock

The rhythms of all organisms are self-sustained, they synchronise with the environment, are subjected to permanent entrainment, but are driven endogenously. The human organism is nowadays considered to be at the crossroads of three different timers: the circadian endogenous clock—the solar clock—the social clock [3, 4].

The circadian clock is a highly convoluted and outstandingly designed network of regulators that interplay throughout 24-hour periods to generate, sustain and synchronise the circadian pattern. This hierarchical biological setup has in its core the central clock that identifies a series of input signals from the environment, having the unique feature of perpetual entrainment, output pathways that interconnect with the endocrine system and the autonomic nervous system, and the most profoundly expressed and intimately embodied circadian machinery, the molecular clock present in every cell. Through a plethora of genes, transcriptional factors, and proteins, this molecular clock undergoes a series of modulations at the level of mRNA and proteins by regulatory transcriptional-translational feedback loops, driven in a tissue-specific manner with the outcome of physiological homeostasis [5, 6].

DOI: http://dx.doi.org/10.5772/intechopen.102021

The solar clock emerges from the Earth's daily rotation, as its surface is sequentially exposed to and respectively deprived of light, basically a geophysical light-dark timetable set by the Sun.

The social clock refers to local time, as it is displayed on a watch and is derived from a combination of solar time and societal responsibilities. Mainly arousing at the age of educational integration of the individual in a strict schedule institution, as it is the case of schools and universities, but also comprising elements regarding feeding times, television and internet spending time, video games for teens, social gatherings with family and friends, the social regulatory pattern is of paramount prominence in understanding the new circadian pattern of modern human organisms [7–9].

4. The circadian topology of humans

The interplay between these three time-setting foundations concealed under a veil of secrecy governs the individual physiologic and behavioural patterns of humankind. As the embodiment of this time-regulatory trinity, humans belong to different chronotypes, depending on their genetic clock mechanisms, as well as their living environment, sex, and age. The circadian topology of humans is mainly depicted by three different characters, the early risers or the morning chronotype, the late risers that express an evening chronotype, and an intermediate or neither type that has rather oscillatory outlines. These typologies are reported to be modulated during the life course, the evening type having a predilection in adolescence and in young adults, a fact that is dramatically changed in later developmental stages, the morning type overlapping to physiological ageing. In addition, gender is quite important in portraying the chronotype, as men are usually reported as being mostly late riser chronotypes. As the rhythmic functioning of the organism is scientifically reasoned, the prevalence of many diseases was causatively linked with circadian disruption. We are nowadays witnessing a tremendous incidence of metabolic disorders, essentially obesity and diabetes, cardiovascular disorders in their outstanding diversity, neurodegenerative impediments, psychiatric imbalances, major sleep conditions, and even cancer and immune system diseases, which are all originating primarily in the misalignment of the inner biochemical circadian state of an individual with the outer environmental liabilities [8–17].

Considering the circadian phenotype, the interindividual differences should be carefully considered when administering drugs, with the particular purpose of augmenting the therapeutic effect and reducing their side effects, in agreement with the 24 hour driven hormones synthesis, blood pressure, body temperature, heart and respiratory rates, central nervous system activity. Clinical chronobiology is a new emerging realm that merges chronopharmacology and chronotherapeutics as areas of medical and pharmaceutical interest, adjusting the treatment time per chronopharmacokinetics and chronopharmacodynamics parameters and the circadian biological unique pattern displayed by a certain patient [18–21].

5. The hierarchical organisation of the human circadian system

In mammals, the central clock, also called pacemaker, that essentially orchestrates the circadian behaviour, resides in the suprachiasmatic nucleus (SCN), located near optic nerves. Even though many theories were issued regarding the mechanisms behind the mammalian circadian clock, today it is generally accepted that after light perception, the SCN sends regulatory outputs toward subsequent CNS levels, and in a cascade of immediate events, it regulates the clock genes expressed in periphery, coordinating the local physiological milieu by rhythmically triggering tissue-specific transcriptional pathways. The hierarchical organisation of this system endorses an orchestrated control imprinted by the central pacemaker, which simultaneously perceives information from the exogenous stimuli and communicates the processed data to downstream effector networks as accurately as cellular mechanisms are envisioned to align the physiological characteristics to the circadian pattern. The endogenous circadian machinery is self-sustained, independent of the presence of external inputs, but in the meantime permanently subjected to alignments by their occurrence, initiating a so-called "photoentrained system" [22–24].

The day-night phase alternation ignites a congregation of signals, mainly light, but also temperature or feeding triggers, which can act as prompts, called *zeitgebers*, meant to synchronise endogenous circadian systems. Fundamentally, the circadian clock is entrained by direct retinal innervation. Light is detected exclusively by the eyes, by a particular arrangement of specific cells and the retinal photopigment, melanopsin. This photic information is conducted by photosensitive ganglion cells expressed at the retina level through the retinohypothalamic tract, directly to the pacemaker, acquiring the entrainment and alignment with the solar clock. Through its rhythmic outputs, the SCN harmonises all the cellular circadian clocks present in every cell of the body organs and tissues to adapt their physiologic features to the external circadian signals [25].

The molecular circadian clock consists briefly of auto-regulated transcriptional and translational feedback loops that are impressed by the oscillatory gene expression controlled by their protein products. The gene expression profiling is under the direct control of the circadian transcriptome, aligning this pattern with the one present in the SCN and peripheral tissues. Circadian transcriptional fluctuations are mandatory for the connected metabolic and functional interplay among various levels of organisation inside a cell, in a particular tissue, and throughout the entire body for the individual roles to be integrated into the complex biological universe of the human organism. A single conductor orchestrates the circadian equilibrium, but every cell has a clear view of the data received from it, not allowing any desynchronisation, nor interferences, in a healthy entity [24, 26].

Several mechanisms may contribute to the complex circadian tissue reprogramming in close relationship with the modulation of the transcription process. The first one is presumably established by epigenetic frames having echoes in the DNA structure that undergoes serious chemical modifications simultaneously with histone proteins. All these amendments command, in their turn, the specific binding of transcription factors to regulatory regions upon the genes, remodelling the transcriptional scenery according to the circadian pattern. The clock genes imprint circadian fluctuations at the epigenome level, correlated with the rhythms exhibited by the transcriptome. The tissue-specific epigenetic arrangements are crucial for setting up cellular identity. In addition, another essential feature regards the interactions established between circadian regulatory proteins and transcriptional factors with characteristic individual profiling, a supplementary regulatory pathway after gene expression regulatory transactions. In line with the previous remarks, the transcriptional expression of regulatory RNA fragments also illustrates a rhythmic profile. Emphasising the cellular machinery synchronisation, the complementary transcriptome regulation through systemic signalling is conducted directly from the SCN or indirectly by the central pacemaker output signals [27, 28].

The SCN is also sensitive to other non-photic inputs from adjacent cerebral regions or the environment, mainly feeding and temperature cues. The feeding time has intense rhythmic reverberations at the level of hepatic transcriptome and the 24-hour body temperature cyclic pattern adjusts the eventually occurring misalignments of the peripheral clock conveyed by various tissues.

Consistent with these clarifications, it should be acknowledged that the health state of an organism is highly dependent on the interplay between circadian molecular and systemic mechanisms, and the extent of its implementation at the peripheral transcriptional level. The most abrupt discrepancies are registered by humans who experience jet lag when travelling between different time zones or the more prevalent case of shift workers. The latter endangers their circadian synchronisation pattern by having a permanently dysregulated light-dark routine and a misaligned feeding schedule. All these aspects are triggering unusual signalling pathways to the inner pacemaker. Consequently, the solar clock and the biological one is completely out of phase, resulting in a disequilibrium generally accredited with the designation "internal desynchronisation" [29–32].

This disturbed state is associated with malfunctioning in transcriptional processes, altered peripheral tissue sensitivity by corrupted input-output transitions, mainly transcribed to chronic disorders located in the most vulnerable territories of the body [33].

To a more profound analysis, the transcriptome disruptions are also of great interest for drugs metabolism, as the cytochrome P450 isoforms and several other enzymes that are relevant for the biologic transformations conducting to toxic metabolites or an amplified structural alteration of the therapeutically active molecules are synthesised based on rhythmically expressed genes. This advances a relevant concern upon the efficacy-toxicity balance of a therapeutic molecule or a xenobiotic, based on its chronopharmacokinetic profile, raising a genuine concern on the time of administration and the specific circadian pattern of a patient (**Figure 1**), which

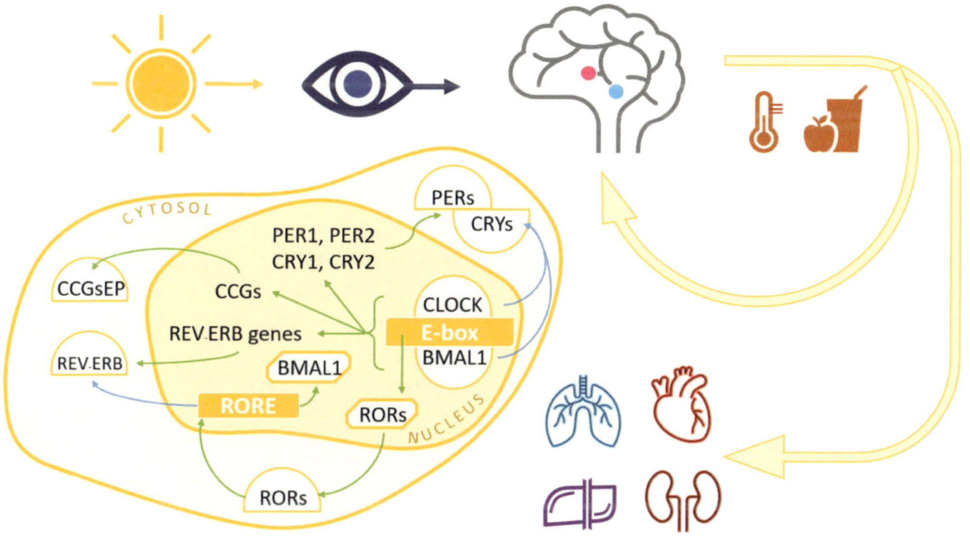

Figure 1.
Global view upon the human circadian clock.

should be considered when special treatment schemes are applied, mainly comprising chemotherapy, cardiovascular and endocrine targeting drugs [34–36].

A significant limitation in designing personalised time-targeted therapies is the absence of relevant biomarkers able to replicate with fidelity the individual circadian features in the highly variable human population. Even though the assessment of a series of hormones levels, namely melatonin, cortisol, DHEA or the body temperature, on a 24-hour basis, can presume roughly the circadian outlines, or that a few transcriptomic and metabolic data can be corroborated and statistically inferred to give a more accurate image of this unique array, no readily available and perfectly concluding techniques can be applied for comprehending the degree of internal desynchronisation. A more desired and attainable perspective would be the actual restoration of the inner state of synchrony by using gradually employed procedures, transcriptomic and environmental triggers to reset timers profoundly endorsed at the molecular level. The outcome should immediately be visible in physiologic sleep-wake cycles, with normally displayed temporal behaviour and the general health state reestablished or at least partially recuperated [37–39].

The circadian machinery is organised in a hierarchy of multiple oscillators, the suprachiasmatic nucleus (SCN) being the central pacemaker at the top of the pyramid. It is synchronised by the 24-hour cycle external signals (the primary input is light and other secondary cues as temperature and feeding), and sequentially, it coordinates the physiological outputs. The multi-oscillator network is synchronised through multiple lines of communication. Peripheral oscillators, present in everybody cell, are reset by timing cues from the SCN, which regulate local circadian physiology. The molecular clock comprises several interconnected transcription feedback loops. The intercellular synchronisation within the SCN is essential for the optimum functioning of the entire body clock.

Author details

Cristina Manuela Drăgoi*, Ion-Bogdan Dumitrescu and Alina Crenguța Nicolae
Faculty of Pharmacy, "Carol Davila" University of Medicine and Pharmacy, Bucharest, Romania

*Address all correspondence to: cristina.dragoi@umfcd.ro

References

[1] de Assis LVM, Oster H. The circadian clock and metabolic homeostasis: Entangled networks. Cellular and Molecular Life Sciences. 2021;**78**:4563-4587. DOI: 10.1007/s00018-021-03800-2

[2] Patke A, Young MW, Axelrod S. Molecular mechanisms and physiological importance of circadian rhythms. Nature Reviews. Molecular Cell Biology. 2020;**21**(2):67-84. DOI: 10.1038/s41580-019-0179-2

[3] Klerman EB, Rahman SA, St Hilaire MA. What time is it? A tale of three clocks, with implications for personalised medicine. Journal of Pineal Research. 2020;**68**(4):e12646. DOI: 10.1111/jpi.12646

[4] Wittmann M, Dinich J, Merrow M, Roenneberg T. Social jetlag: Misalignment of biological and social time. Chronobiology International. 2006;**23**(1-2):497-509. DOI: 10.1080/07420520500545979

[5] Archer SN et al. Phenotyping of PER3 variants reveals widespread effects on circadian preference, sleep regulation, and health. Sleep Medicine Reviews. 2018;**40**:109-126

[6] Phillips AJK et al. High sensitivity and interindividual variability in the response of the human circadian system to evening light. Proceedings of the National Academy of Sciences of the United States of America. 2019;**116**(24):12019-12024

[7] Mota MC, Silva CM, Balieiro LCT, Gonçalves BF, Fahmy WM, Crispim CA. Association between social jetlag food consumption and meal times in patients with obesity-related chronic diseases. PLoS One. 2019;**14**(2):e0212126. DOI: 10.1371/journal.pone.0212126

[8] Roenneberg T, Winnebeck EC, Klerman EB. Daylight saving time and artificial time zones—A battle between biological and social times. Frontiers in Physiology. 2019;**10**(944):1-12

[9] Roenneberg T, Pilz LK, Zerbini G, Winnebeck EC. Chronotype and social jetlag: A (self-) critical review. Biology. 2019;**8**(3):54. DOI: 10.3390/biology 8030054

[10] Hsu PK, Ptacek LJ, Fu YH. Genetics of human sleep behavioral phenotypes. Methods in Enzymology. 2015;**552**:309-324

[11] Papantoniou K, Pozo OJ, Espinosa A, Marcos J, Castaño-Vinyals G, Basagaña X, et al. Increased and mistimed sex hormone production in night shift workers. Cancer Epidemiology, Biomarkers & Prevention. 2015;**24**:854-863

[12] Rijo-Ferreira F, Takahashi JS. Genomics of circadian rhythms in health and disease. Genome Medicine. 2019;**11**:82. DOI: 10.1186/s13073-019-0704-0

[13] Roenneberg T, Wirz-Justice A, Merrow M. Life between clocks: Daily temporal patterns of human chronotypes. Journal of Biological Rhythms. 2003;**18**:80-90

[14] Roenneberg T, Daan S, Merrow M. The art of entrainment. Journal of Biological Rhythms. 2003;**18**:183-194. DOI: 10.1177/0748730403018003001

[15] Rybak YE, McNeely HE, Mackenzie BE, Jain UR, Levitan RD. An open trial of light therapy in adult attention-deficit/hyperactivity disorder.

Journal of Clinical Psychiatry. 2006; **67**(10):1527-1535. DOI: 10.4088/JCP. v67n1006

[16] Sit DK, McGowan J, Wiltrout C, Diler RS, Dills JJ, Luther J, et al. Adjunctive bright light therapy for bipolar depression: A randomized double-blind placebo-controlled trial. American Journal of Psychiatry. 2018;**175**(2):131-139. DOI: 10.1176/appi.ajp.2017.16101200

[17] Voiculescu SE, Le Duc D, Roşca AE, Zeca V, Chiţimuş DM, Arsene AL, et al. Behavioral and molecular effects of prenatal continuous light exposure in the adult rat. Brain Research. 1650;**2016**:51-59

[18] Adam D. Core concept: Emerging science of chronotherapy offers immense opportunities to optimise drug delivery. Proceedings of the National Academy of Sciences of the United States of America. 2019;**116**(44):21957-21959. DOI: 10.1073/pnas.1916118116

[19] Dong D, Yang D, Lin L, Wang S, Wu B. Circadian rhythm in pharmacokinetics and its relevance to chronotherapy. Biochemical Pharmacology. 2020;**178**:114045. DOI: 10.1016/j.bcp.2020.114045

[20] Lu D, Zhao M, Chen M, Wu B. Circadian clock-controlled drug metabolism. Implications for chronotherapeutics. Drug Metabolism and Disposition. 2020;**48**(5):395-406. DOI: 10.1124/dmd.120.090472

[21] Ruben MD, Smith DF, FitzGerald GA, Hogenesch JB. Dosing time matters. Science. 2019;**365**(6453):547-549. DOI: 10.1126/science.aax7621

[22] Ayyar VS, Sukumaran S. Circadian rhythms: Influence on physiology, pharmacology, and therapeutic interventions. Journal of Pharmacokinetics and Pharmacodynamics. 2021;**48**:321-338. DOI: 10.1007/s10928-021-09751-2

[23] Laje R, Agostino PV, Golombek DA. The times of our lives: Interaction among different biological periodicities. Frontiers in Integrative Neuroscience. 2018;**12**:1-6. DOI: 10.3389/fnint.2018.00010

[24] Wright KP Jr et al. Entrainment of the human circadian clock to the natural light-dark cycle. Current Biology. 2013;**23**(16):1554-1558

[25] Fernandez DC, Chang YT, Hattar S, Chen SK. Architecture of retinal projections to the central circadian pacemaker. Proceedings of the National Academy of Sciences. 2016;**113**(21):6047-6052. DOI: 10.1073/pnas.1523629113

[26] Wirz-Justice A, Skene DJ, Münch M. The relevance of daylight for humans. Biochemical Pharmacology. 2020;**191**:114304. DOI: 10.1016/j.bcp.2020.114304

[27] Dragoi CM, Nicolae AC, Dumitrescu IB, Popa DE, Ritivoiu M, Arsene AL. DNA targeting as a molecular mechanism underlying endogenous indoles biological effects. Farmácia. 2019;**67**:367-377

[28] Ruben MD, Wu G, Smith DF, Schmidt RE, Francey LJ, Lee YY, et al. A database of tissue-specific rhythmically expressed human genes has potential applications in circadian medicine. Science Translational Medicine. 2018;**10**(458):eaat8806. DOI: 10.1126/scitranslmed.aat8806

[29] Dragoi CM, Morosan E, Dumitrescu IB, Nicolae AC, Arsene AL, Draganescu D, et al. Insights into chrononutrition: The innermost interplay

amongst nutrition, metabolism and the circadian clock, in the context of epigenetic reprogramming. Farmácia. 2019;**67**:557-571

[30] Kecklund G, Axelsson J. Health consequences of shift work and insufficient sleep. BMJ. 2016;**355**:i5210

[31] Maury E, Ramsey KM, Bass J. Circadian rhythms and metabolic syndrome: From experimental genetics to human disease. Circulation Research. 2010;**106**(3):447-462. DOI: 10.1161/CIRCRESAHA.109.208355

[32] Nicolae AC, Dragoi CM, Ceausu I, Poalelungi C, Iliescu D, Arsene AL. Clinical implications of the indolergic system and oxidative stress in physiological gestational homeostasis. Farmácia. 2015;**63**:46-51

[33] Depner CM, Stothard ER, Wright KP Jr. Metabolic consequences of sleep and circadian disorders. Current Diabetes Reports. 2014;**14**(7):507

[34] Brainard J, Gobel M, Scott B, Koeppen M, Eckle T. Health implications of disrupted circadian rhythms and the potential for daylight as therapy. Anesthesiology. 2015;**122**(5):1170-1175. DOI: 10.1097/ALN.0000000000000596

[35] Hermida RC, Crespo JJ, Domínguez-Sardiña M, Otero A, Moyá A, Ríos MT, et al. Bedtime hypertension treatment improves cardiovascular risk reduction: The Hygia chronotherapy trial. European Heart Journal. 2020;**41**(48):4565-4576. DOI: 10.1093/eurheartj/ehz754

[36] Levi F et al. Implications of circadian clocks for the rhythmic delivery of cancer therapeutics. Advanced Drug Delivery Reviews. 2007;**59**(9-10):1015-1035

[37] Boghitoiu D, Grama A, Pop T, Simionescu A, Ghita I, Ulmeanu CE, et al. The role of microRNAs as diagnostic biomarkers in the early prediction of acetaminophen-induced liver injury. Farmácia. 2021;**69**:4. DOI: 10.31925/farmacia.2021.4.21

[38] Masri S, Sassone-Corsi P. The emerging link between cancer, metabolism, and circadian rhythms. Nature Medicine. 2018;**24**(12):1795-1803. DOI: 10.1038/s41591-018-0271-8

[39] Sen A, Hoffmann HM. Role of core circadian clock genes in hormone release and target tissue sensitivity in the reproductive axis. Molecular and Cellular Endocrinology. 2020;**501**:110655. DOI: 10.1016/j.mce.2019.110655

Chapter 2

Circadian Sensation and Visual Perception

Michael Jackson Oliveira de Andrade

Abstract

The physiology of living beings presents oscillations that are known as biological rhythms. The most studied rhythm is called circadian (circa = circa, dies = day), because it varies with a period close to 24h. Most functions of the body have circadian variations, one can mention, for example, metabolism, body temperature, the activity of the nervous system, secretion of hormones such as melatonin and cortisol. Circadian rhythms were also found in human behavior, for example: in sensory activity, motor activity, reaction time, visual perception, auditory perception, time perception, attention, memory, arithmetic calculus, and executive functions. The present work reviews the visual path that participates in the synchronization of circadian rhythms, as well as the evidence that exists about the presence of circadian rhythms in the sensation and visual perception of the human being.

Keywords: circadian rhythms, sensation, visual perception

1. Introduction

1.1 Circadian rhythms

Circadian rhythms have several characteristics that are important to analyze. Under constant environmental conditions, these rhythms persist in a period close to 24 h. However, some of the physiological functions have different periods, so that circadian rhythm are considered as part of a multioscillative system [1]. However, the circadian rhythms of the different functions remain synchronized by a central pacemaker, located in the suprachiasmatic nucleus of the hypothalamus [2]. A set of genes involved in the generation and modulation of these rhythms was verified [3]. It was found that some environmental events keep circadian rhythms synchronized with environmental cycles. These synchronization agents are lighting cycles, room temperature, food availability, exercise, and social stimulation. Ambient lighting cycles are the most effective synchronize circadian rhythm, both in animals and in humans [4].

Three neurophysiological pathways participate in the responses of humans to light. The first clue includes two types of receptors in the retina, rods and cones, which respond to the intensity and frequency of light. These receptors are connected with bipolar and amharic cells, and ganglion cells that connect to neurons in the lateral geniculate nucleus of the thalamus and neurons in the occipital cortex. This nervous path participates in the analysis of shapes, colors, and images. The second pathway involves ganglion cells in visual connections with nuclei of the superior colliculus of the thalamus midbrain, pulvinar nuclei, and the inferior temporal cortex. This nervous path participates in the identification of space. The third visual pathway includes a group of retinal ganglion cells that respond to light, have connections to the suprachiasmatic nucleus and participate in the mechanism of synchronization of circadian rhythms. These retinal neurons are known as intrinsically photosensitive retinal ganglion cells (ipRGCs), retinal ganglion cells give photosensitive (CGRP), or retinal ganglion cells that contain melanopsin (mRGCs) [5, 6].

2. Visual pathway that participates in the synchronization of circadian rhythms

Photosensitive ganglion cells are located in greater quantity in the center of the retina with a decrease towards the periphery, their axons protrude into the suprachiasmatic nucleus and other subcortical visual areas involved in the entrainment of light Dacey et al. [7]. It is important to note that only 3% of the total ganglion cell population is photosensitive. Morphologically ipRGCs present large receptive fields and their depolarization (inverse process of rod neurons and hyperpolarized cones) happens in dendrites and axons. They can be classified into five cell subtypes [8]: cells type M1 (have higher neuronal density and higher levels of opsin expression), cells type M2 and M3, (have levels of expression of medium opsin), and cells of type M4 and M5 (have low levels of expression of opsin). Ecker et al. [9] showed that M2 and M4 cells send neural information to various regions of the brain, including areas involved in visual processing through dorsal and ventral areas of the lateral geniculate nucleus of the hypothalamus.

The dendrites of ipRGCs receive photoreceptor inputs (cones and rods) via bipolar and amharic cell connections [10]. These findings hypothesized that ipRGC cells relay luminosity-sensitive visual input signals in sustained circadian time. Since these pathways integrate with the nervous system responsible for visual processing, it is possible that some ipRGCs axons carry output signals derived from various photopigments [11].

Provencio et al. [5] identified and defined a model of regulation of circadian rhythms through ipRGCs. Initially, this model discusses that light excites a group of photosensitive opsins in ipRGC s cells and later induces the opening of glutamatergic receptors in the suprachiasmatic nucleus. At complex levels, a lasting expression of genes occurs by the joining of some proteins that fit in a period of approximate de 24 h [6, 12, 13].

IpRGCs contain opsins called melanopsins act as light-sensitive photopigments [14, 15]. Melanopsin has signaling time properties in ipRGCs distinct from cones and rods [7, 16, 17], being associated with circadian, neuroendocrine and neurobehavioral functions, besides influencing some imaging functions [18]. Electrophysiological records of ipRGCs have shown lower sensitivity to light in relation to photoreceptor cones and rods [13]. According to the standard view, this photopigment has low sensitivity to light but is able to integrate sustained photic energy.

The peak visual sensitivity of ipRGCs (420 to 480 nm) crosses with the light intensity absorbed by rods and cones (380–650 nm and 440–560 nm, respectively), mainly related to short wavelength and intensity 430 nm (**Figure 1**). Melanopsin adaptation contributes to the relative gain of spectral sensitivity of photoreceptor responses

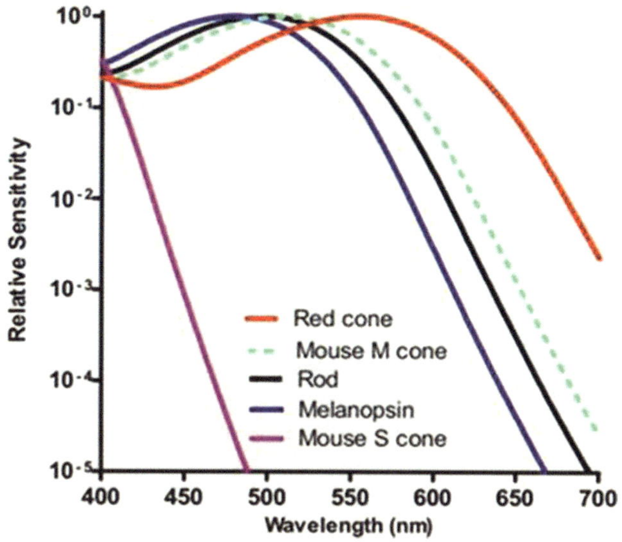

Figure 1.
Discovery of ipRGCs: Approximate spectral sensitivity by opsin nanogram with peak human sensitivity for long-length cones (red line, lmax = 556 nm), peak spectral sensitivity of a rod rat (black line, lmax = 498 nm), melanopsin (blueline, lmax = 480 nm), and cones of short wavelength (purple line, lmax = 360 nm) and medium (dotted line, lmax = 511 nm). Adapted from [19].

to cones and rods in the retina [20]. However, the individual contribution of each class of photoreceptors to irradiance responses is complex [19]. That is, the circadian rhythm system is part of a complex visual system [21].

According to Berson (2003) the photosensitivity of ipRGCs requiring melanopsin conduct various visual functions in the absence of rods and cones. Studies by Tosini [22] and Tosini, Pozdeyev, Sakamoto and Luvone [23] suggest that the retina also has a circadian clock capable of controlling visual processing functions. Numerous studies have shown that visual sensitivity, defined by means of visual contrast thresholds, has a circadian behavior in several species, among them: zebras [24], Larval Xenopus [25], rats [26], and humans [27–30]. These results, in general, culminate the idea that the filtering of properties of human visual processing occurs according to the circadian variation of visual sensitivity. Therefore, ipRGCs cells constitute a third class of photoreceptors, in addition to rod cells and cones, which are responsible for the formation of the image or visual scene during a 24-hour rhythm.

Remember that rods and cones ingrate ipRGCs and, consequently, circadian behaviors may be associated with these mechanisms. Recent findings have shown that mechanisms that process the contrast of luminance and color are associated with the circadian timing pattern [27, 31].

3. Circadian rhythms of sensation and visual perception

Circadian rhythms modulate the physiology and behavior of animals and humans, so it is important to analyze which aspects of visual perception present circadian variations. Next, we describe the evidence that exists about circadian rhythms in the sensation and perception of contrast, as well as the sensation and perception of the contrast of luminance and colors.

3.1 Daily variation of sensation and perception of contrast of luminance

Melanopsin can contribute by balancing spatial and temporal resolution and optimizing visual sensitivity performance, even at low response levels [32]. Studies on the relationship between circadian mechanisms and visual sensitivity of luminance are still considered contradictory. However, electrophysiological and psychophysical studies have raised preliminary hypotheses about the circadian influence on the discrimination of the visual threshold of luminance contrast. Spatial variation in light intensity, called spatial contrast, comprises much of the visual information perceived by mammals, and the relative ability to detect contrast is called contrast sensitivity [33].

As previously seen, Turner and Mainster [34] argue that retinal ganglion cell photopigments contribute to the circadian modulation of visual sensory function [35, 36]. O'Keefe and Baker [37] used the psychophysical method of constant stimulus to measure diurnal variations in visual sensitivity in photon and scotopic luminance conditions. This study showed higher visual sensitivity in photon conditions compared to scotopic sensitivity. In addition, the visual sensitivity of the individuals varied according to the daily time, presenting greater visual sensitivity at night. Using a similar psychophysical method, Tassi et al. [30] suggested that human sensitivity changes during the 24-hour period and that circadian rhythm has an influence on visual sensitivity in both scotopic and mesopic luminance conditions (0.007 and 0.021 cd/m^2).

Tassi et al. [30] used a daily routine chronobiological protocol that provides a schematic representation of the subject's responses during the peaks inactivity. The authors scored eight measurement points with alternation of 2 h during the period from 08:00 to 20:00 hours, and showed changes in visual sensitivity throughout the day, with decreased sensitivity in the morning and progressive increase throughout the day, reaching its peak at 22:00. It was also observed that the sensitivity remained high during the first half of the night and progressively decreased after 04:00 in the morning.

Bassi and Powers [28] argued that temporal visual sensitivity also exhibits variation during the day. The authors measured the visual thresholds of seven subjects between 20 and 38 years old using the psychophysical paradigm of equalization by flashing photometry (flicker) during the periods of (12:00–14:00 and 00:00–02:00). Bassi and Powers noticed a small fluctuation in visual sensitivity and concluded that the ability to detect light varied systematically according to the periods indicated, showing minimal sensitivity during 12:00–14:00 and presenting maximum sensitivity at 00:00–02:00.

Although the findings indicated visual circadian alterations, the studies did not pay attention to verifying whether this variation differed according to the circadian typology of the participants, that is, whether the circadian modulation of visual sensitivity has different characteristics between subjects with behavioral patterns of morning activity and rest, intermediates and afternoons. In 1997 Tassi and Pins used the Horne and Ostberg Questionnaire to characterize the circadian typology of seven subjects and measured binocular visual sensitivity with a psychophysical method of adaptation between 08:00 and 18:00. The authors observed that visual sensitivity was low during the morning shift and increased progressively from 10:00 am, remaining constant until 18:00 hours.

The results for low sensitivity in the morning shift were only for three subjects with intermediate chronotype, while the sensitivity that remained constant was for the subjects with morning and afternoon chronotype. Still, Tassi and Pins argued that visual reaction time is significantly longer during the night period, and the proportion of responses between the sexes indicates that men have greater stability than women. In general, the authors suggest that this phenomenon of sensitivity variation is not a general rule since subgroups could be isolated in samples considering their circadian typology, in addition several mechanisms can contribute to fluctuation of visual sensitivity, such as the rhythmic expression of melatonin and dopamine in rods and photoreceptors and cones.

3.2 Daily variation in contrast sensitivity

The ability to detect and interpret details of a visual scene is determined by the visual system's ability to distinguish contrast patterns [38]. Contrast can also be understood as the physical property of the visual stimulus, and the magnitude of the variation of luminance in the stimulus related to the total luminance of adjacent areas [38].

In this sense, the visual system (SV) has high sensitivity when a pattern needs little contrast to be detected. The reverse, low sensitivity, when the SV needs high contrast value to detect the stimulus. The contrast sensitivity curve represents the sensitivity of the SV in being more or less sensitive at certain spatial frequencies [33, 39]. The frequency of a sine wave in a visual experiment is described by visual angle degree cycles, which corresponds to the number of grid cycles that subtend 1 degree of angle to the human eye [40]. The Contrast Sensitivity Function (CSV) of numerous thresholds describes an inverted U-shaped curve with low frequency inclination and a steep slope in high frequencies [38]. From a practical point of view, CSV can be measured through psychophysical or behavioral criteria [41]. Physiologically, these channels refer to neuronal populations involved in the selective processing of spatial frequency bands [42]. Thus, it is believed that the process of detecting visual contrast is due to the activity of one or more spatial frequency channels, the variations of these frequencies are broken down into bands tuned to low, medium and high spatial frequencies.

Thus, it is possible to discriminate the minimum amount of contrast that the SV needs to individually detect each range or spatial frequency band since the response of one channel is not affected by the response of the other. So far, the big question discussed is whether there is a circadian rhythm of CSV, that is, that there is a circadian SV of contrast sensitivity of spatial frequencies.

Struck, Rodnitzky, and Dobson [43] used sine wave grid stimuli to measure CSV. Struck et al. [43] evaluated the contrast sensitivity of 12 participants with Parkinson's disease (PD) using low, medium and high spatial frequencies (1, 5, 3, 6, 12 and 18 cpg). The authors found punctual circadian fluctuations with lower dysfunction in the early morning compared to the afternoon. At 8:30 a.m., sensitivity in participants did not differ from healthy subjects, but as hours passed, the results were significantly worse at 3 or more spatial frequencies (3 and 6 cpg). CSV for healthy subjects remained constant over time, unlike the findings on visual circadian fluctuations found by Bassi and Powers [28], O'Keefe and Baker [37], Tassi and Pins [29], and Tassi et al. [30]. For Struck et al. [43], the variation in contrast sensitivity in spatial frequencies was related to dopamine deficiency caused by PD.

Recent studies prepared by Andrade, Silva, and Santos [44] and Andrade et al. [27, 31] evaluated CSV circadian fluctuation in healthy adult subjects according to circadian typology at different times of the day. These studies indicate that CSV curves have minimum sensitivity at 09:00, progressive increase from 13:00 to 17:00 and maximum sensitivity at 21:00. During the periods of 13:00 and 17:00 the measurements remain constant. Also, morning subjects present peak sensitivity in the morning, but are less sensitive than afternoon subjects throughout the day.

Andrade et al. [44] measured the CSV of 18 male subjects using the daily routine chronobiological protocol and the psychophysical method of the ladder to measure stimuli of vertical sine grid in a condition of photopic (41.9 cd/m^2). The results showed that morning subjects have maximum sensitivity at 7:00 a.m. when compared to the 3:00 and 11:00 times, mainly at the spatial frequency of 3.1 cpg; intermediate and vespertine subjects presented maximum sensitivity at 23:00 for spatial frequencies of 0.6 and 3.1, 6.1 and 8.8 cpg, respectively. It is possible to observe a fluctuation in CSV as seen by Bassi and Powers [28]. Thus, in addition to the variation in daily luminance contrast, the findings also indicated differences in sensitivity according to circadian typology in relation to specific spatial frequencies. This study demonstrated that visual contrast adaptation depends on behavioral and cognitive factors such as sleep latency, drowsiness levels, alertness and visual attention. From this point, Andrade et al. [44] highlight the needs of psychophysical research that evaluates basic visual functions in terms of chronobiological rhythms, such as the measurement of CSV in circadian time [45].

Andrade et al. [27, 31] also evaluated CSV according to circadian typology and measurement time of adult men ($M = 23.42 \pm 2.6$ years) during a daily period. The authors observed that CSV curves had decreased sensitivity for low and high spatial frequencies [46] and peak sensitivity for average spatial frequencies [42]. The results did not show variation in CSV of morning subjects, but detected fluctuation in CSV for night subjects with greater sensitivity in the period of 21:00 in spatial frequencies of 1.0, 3.1 and 13.2 cpg, and fluctuation in CSV for intermediate subjects in spatial frequencies of 0.2 and 15.6 cpg. Furthermore, the study by Andrade et al. [27, 31] pointed out that intermediate subjects have maximum sensitivity in spatial frequency 0.2 cpg at 17:00 when compared to 9:00 and 13:00 times. There was an attenuation in visual sensitivity during the morning in all groups, except for morning subjects. In general, the study by Andrade et al. [27, 31] proposes greater visual sensitivity in the CSV curve during the night period (21:00) and a decrease in sensitivity at 09:00. According to Viola et al. [47] it is common to observe greater variation in behavioral measures of venous subjects due to evolutionary traits and circadian genetic expression of these subjects, i.e., venous subjects have a genetic repeat polymorphism $PER3^{4/4}$ more vulnerable to daily fluctuation, when compared to the genetic expression of morning subjects $PER3^{5/5}$ and intermediate subjects $PER3^{4/5}$. Furthermore, evening subjects have a tendency to sleep delay (2 h) in relation to other subjects, thus night vision adapts melatonin production and masks the significant effect of spatial resolution of photoreceptors rods and cones [48].

Both results described by Andrade et al. [44] indicate maximum sensitivity at 07:00/09:00 compared to 15:00 and 23:00 for morning subjects and maximum sensitivity in the periods of 15:00 and 23:00 for intermediate and vespertine subjects, characterizing possible patterns of behavior of CSV according to circadian typology.

Daily changes in visual sensitivity can explain many factors about the effects of light on visual contrast perception. In fact, the visual system is not linear and the visual sensitivity curve presents daily variations according to visual space and time [44], however these variations may be related to circadian behavior of CSV (**Figure 2**). Thus, the relationship between homeostatic and circadian variables establishes a daily control of the detection of visual thresholds of luminance contrast [27, 31].

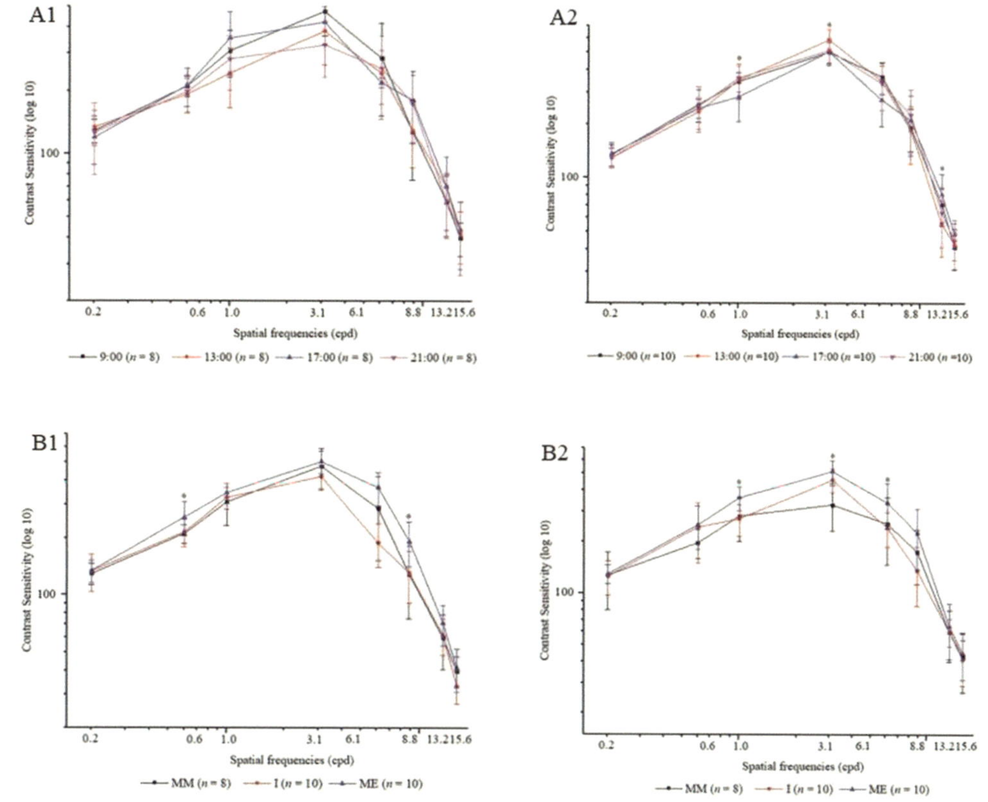

Figure 2.
Circadian behavior of the visual contrast curves: (A) contrast sensitivity curves obtained during the 24-hour period according to the time: (A1) represents the variation of the contrast in spatial frequencies for the moderately morning participants (MM); represents the variation of the contrast in the spatial frequencies for the participants of the moderately afternoon (MV); (B) contrast sensitivity curves obtained during the 24-hour period for participants MM, MV and intermediates (I): (B1) represents the variation of contrast in spatial frequencies at 9 am; (B2) represents the variation of contrast in spatial frequencies at 9 pm. Taken from Andrade et al., [27, 31].

3.3 Daily variation of sensation and perception of colors

In recent decades, the study of color perception has stood out due to its functional role in the visual perception process of subjects with normal vision and deviations in the axes of color confusion. In general, the human view of color is considered trichromatic and its modification results from congenital defects or is acquired by adverse situations [49]. A set of photoreceptor cells in several visual sensitivity levels can absorb and combine wavelengths to form three-dimensional color perception at any point in the visual field [50, 51].

The variability of light absorption in the retina by cones shows that opsins are sensitive to wavelengths in the range of 400 to 700 nm; their absorption intensity varies according to the selectivity of the three classes or wavelength peaks, that is, to red (559 nm), green (531 nm) and blue (419 nm). It is appropriate to refer to the three types of cone as long, medium and short wave, respectively. According to Figueiro, Bullough, Parsons and Rea [52], the physiological aspects of retinal imaging are still subject to a biological change in time caused by circadian rhythm. Psychophysical studies show that color perception can be evaluated based on the capacity of equality and discrimination of thresholds [51, 53]. Studies have investigated the circadian mechanisms of color perception [54, 55].

Pauers et al. [55] suggest that the color opposition mechanism is evolutionarily adapted to the spectral changes of sunlight during the earth's rotation. In addition, the spectral positioning of the short wavelength S (absorption for blue color perception) may be directly related to the determination of these patterns of activity.

According to a study conducted by Danilenko et al. [56], a greater capacity of activity of cones during the night and a decrease in response in the early morning was reported through an electroretinogram during a constant follow-up of 24 hours of white light. The data indicate that there is a change in the perception of color during a circadian rhythm, but it is not clear in relation to the period, amplitude and phase in which it occurs. For Ebihara and Tsuji [57] and Mrosovsky [58], melanopsin also contributes to the synchronization of circadian rhythm activity of visual color. However, it is possible that a number of behavioral and physical variables, such as the light/dark cycle, exposure to light and circadian typology determine variations in the axes of color confusion [59]. Thus, studies start from hypotheses that there is a daily modulation of visual color perception in the long, medium and short wavelength confusion axes [31].

Andrade et al. [27, 31] measured the circadian rhythm of the color confusion axes of 28 young male adults aged between 20 and 28 years according to their circadian typology using the Cambridge Color Test (CCT) and the Lanthony Desaturated D-15d test). This study suggested that the neural processing of color perception in the green-red axis has a daily fluctuation, especially for night subjects. In addition, the vespertine subjects presented higher visual sensitivity in the protan, deutan and tritan axes, especially in the periods of 9:00 and 21:00 and morning subjects had lower visual sensitivity at 13:00 and 17:00. To Archer et al. [60], morning subjects are more stable in relation to fluctuations in behavioral responses. It is normal that there is a variation of the visual thresholds during the day, and that they present normal parameters for trichromatic subjects [61], however it is possible that this variation has characteristics according to circadian typology and measurement times.

The modulation in the operation of the green-red and blue-yellow confounding axes may be related to the interactions of wavelength absorption by ipRGC cells (420 to 480 nm) that intersect with the light intensity absorbed by the rods and cones (380-650 nm and 440–560 nm, respectively). Thus, Tritan axis stability is intrinsically related to the constant absorption of melanopsin and stability of the daily fluctuation of the blue wavelength [22, 62, 63]. Danilenko et al. [56] suggest that the spectral position of the medium and long wavelength suffers less influence of melanopsin pigment activity, allowing the daily fluctuation of chromatic sensitivity. According to Walmsley et al. [64] the variation of these spectral changes is necessary for circadian alignment of information, keeping natural conditions constant. It is a note point that the daily entrainment of sleep capacity can contribute significantly to the daily adaptation of visual perception [27, 31].

Thus, it is possible to characterize chromatic rhythmic changes of the visual system according to exposure to light, however the results for color perception are considered preliminary [65].

4. Final considerations

In the present work, we reviewed the theoretical and experimental knowledge discovered about the interaction between the visual system and the circadian system. A visual path was found that participates in the synchronization of circadian rhythms. Light produces synchronization of circadian rhythms by activating intrinsically photosensitive retinal ganglion cells (ipRGCs). These cells use melanopsin as a photopigment and modulate the activity of the suprachiasmatic nucleus of the hypothalamus, which acts as a central pacemaker of circadian rhythms. The analysis of the functioning of the visual path that synchronizes circadian rhythms is important to understand how the human being adapts to schedule changes, such as transnational air travel, night work, or rotating shifts. This visual path is also relevant to understanding changes in response to light, which have been observed in patients with circadian rhythm disorders, such as: circadian rhythm sleep disorder, seasonal depression or manic-depressive disorder. In addition, knowledge of this visual pathway is important to analyze the physiological effects of light (phototherapy), when used as a treatment for circadian rhythm disorders, as well as other neurological and psychiatric disorders.

On the other hand, circadian rhythms were found in submodality of visual sensation and perception, such as contrast of luminosity, color discrimination and perception of objects, and images. These rhythms in visual perception are modulated by people's chronotypes. The study of circadian rhythms in visual perception can be important for the learning of visual and visuospatial tasks, as well as for performance in school or work activities that require more specific visual processes, such as contrast of magnifying detail or color perception. In the meantime, it is also important to analyze the interaction of circadian rhythms in visual perception with rhythms of other cognitive processes, such as attention, memory and executive functions; as well as the role of these rhythms in work performance.

Author details

Michael Jackson Oliveira de Andrade
Laboratory of Neuroscience, Chronobiology and Sleep Psychology, Department of Psychology, State University of Minas Gerais, MG, Brazil

*Address all correspondence to: andrademjackson@gmail.com

References

[1] Bell-Pedersen D, Cassone VM, Earnest DJ, Golden SS, Hardin PE, Thomas TL, et al. Circadian rhythms from multiple oscillators: Lessons from diverse organisms. Nature Reviews Genetics. 2005;**6**:544-556

[2] Evans J, Silver R. The suprachiasmatic nucleus and the circadian timekeeping system of the body. In: Pfaff D, Volkov N, editors. Neuroscience in the 21st Century. New York, NY: Springer; 2016. pp. 2241-2288

[3] Albrecht U, Eichele G. The mammalian circadian clock. Current Opinion in Genetics & Development. 2003;**13**(3):271-277. DOI: 10.1016/s0959-437x(03)00055-8

[4] Czeisler CA, Buxton OM. Human circadian timing system and sleep-wake regulation. In: Kryger MH, Roth T, Dement WC, editors. Principles and Practice of Sleep Medicine. 6th ed. Philadelphia, PA: Elsevier Saunders; 2017. pp. 362-376

[5] Provencio I, Jiang G, De Grip WJ, Hayes WP, Rollag MD. Melanopsin: An opsin in melanophores, brain, and eye. Proceedings of the National Academy of Sciences of the United States of America. 1998;**95**(1):340-345

[6] Sexton T, Buhr E, Van Gelder RN. Melanopsin and mechanisms of non-visual ocular photoreception. Journal Biology Chemical. 2012;**287**(3):1649-1656. DOI: 10.1074/jbc. R111.301226

[7] Dacey M, Liao HW, Peterson BB, Robinson FR, Smith VC, Pokorny J, et al. Melanopsin-expressing ganglion cells in primate retina signal colour and irradiance and project to the LGN. Nature. 2005;**433**:749-754. DOI: 10.1038/nature03387

[8] Brainard GC, Hanifin JP, Greeson JM, Byrne B, Glickman G, Edward G, et al. Action spectrum for melatonin regulation in humans: Evidence for a novel circadian photoreceptor. The Journal of Neuroscience. 2001;**21**(16):6405-6412

[9] Ecker JL, Dumitrescu ON, Wong KY, Alam NM, Chen SK, LeGates T, et al. Melanopsin-expressing retinal ganglion-cell photoreceptors: cellular diversity and role in pattern vision. Neuron. 2010;**67**(1):49-60. DOI: 10.1016/j.neuron.2010.05.023

[10] Hankins MW, Peirson SN, Foster RG. Melanopsin: an exciting photopigment. Trends Neuroscience. 2008;**31**(1):27-36. DOI: 10.1016/j.tins.2007.11.002

[11] Berlinky MA, Smerasky CA, Provencio I, Sollars PJ, Pickard GE. Melanopsin retinal ganglion cells receive bipolar and amacrine cell synapses. Journal Comp Neurology. 2003;**460**(3):380-393

[12] Reppert SM, Weaver DR. Molecular analysis of mammalian circadian rhythms. Annual Review of Physiology. 2001;**63**:647-676. DOI: 10.1146/annurev.physiol.63.1.647

[13] Sand A, Schmidt TM, Kofuji P. Diverse types of ganglion cell photoreceptors in the mammalian retina. Progress in Retinal and Eye Research. 2012;**31**(4):287-302. DOI: 10.1016/j.preteyeres.2012.03.003

[14] Provencio I, Rodriguez IR, Jiang G, Hayes WP, Moreira EF, Rollag MD. A novel human opsin in the inner retina. The Journal of Neuroscience. 2000;**20**(2):600-605

[15] Rea MS, Figueiro MG, Bierman A, Bullough JD. Circadian light. Journal of

Circadian Rhythms. 2010;**8**(2):2-10. DOI: 10.1186/1740-3391-8-2

[16] Joyce DS, Feig B, Cao D, Zele AJ. Temporal characteristics of melanopsin inputs to the human pupil light reflex. Vision Research. 2015;**107**:58-66. DOI: 10.1016/j.visres.2014.12.001

[17] Lucas RJ, Lall GS, Allen AE, Brown TM. How rod, cone, and melanopsin photoreceptors come together to enlighten the mammalian circadian clock. Progress in Brain Research. 2012;**199**:1-18. DOI: 10.1016/B978-0-444-59427-3.00001-0

[18] Feigl B, Zele AJ. Melanopsin-expressing intrinsically photosensitive retinal ganglion cells in retinal disease. Optom Visual Science. 2014;**91**(8):894-903. DOI: 10.1097/OPX.0000000000000284

[19] Lall GS, Revell VL, Momiji H, Al Enezi J, Altimus CM, Guller AD, et al. Distinct contributions of rod, cone, and melanopsin photoreceptors to encoding irradiance. Neuron. 2010;**66**(3):417-428. DOI: 10.1016/j.neuron.2010.04.037

[20] McDougall DH, Gamlin PD. The influence of intrinsically photosensitive retinal ganglion cells on the spectral sensitivity and response dynamics of the human pupillary light reflex. Vision Research. 2010;**50**(1):72-87. DOI: 10.1016/j.visres.2009.10.012

[21] Morin LP, Allen CN. The circadian visual system. Brain Research Reviews. 2005;**51**(1):1-60. DOI: 10.1016/0165-0173(94)90005-1

[22] Tosini G. Melatonin circadian rhythm in the retina of mammals. Chronobiology International. 2000;**17**(5):599-612

[23] Tosini G, Pozdeyev N, Sakamoto K, Luvone PM. The circadian clock system in the mammalian retina. BioEssays. 2008;**30**(7):624-633

[24] Li P, Temple S, Gao Y, Haimberger TJ, Hawryshyn CJ, Li L. Circadian rhythms of behavioral cone sensitivity and long wavelength opsin mRNA expression: a correlation study in zebrafish. The Journal of Experimental Biology. 2005;**208**(3):497-504. DOI: 10.1242/jeb.01424

[25] Solessio E, Scheraga D, Engbretson GA, Knox BE, Barlow RB. Circadian modulation of temporal properties of the rod pathway in larval Xenopus. Journal of Neurophysiology. 2004;**92**(5):2672-2684. DOI: 10.1152/jn.00344.2004

[26] Hwang CK, Chaurasia SS, Jackson CR, Chan GCK, Storm DR, Iuvone PM. Circadian rhythm of contrast sensitivity is regulated by a dopamine–neuronal PAS-domain protein 2–adenylyl cyclase 1 signaling pathway in retinal ganglion cells. Journal Neuroscience. 2013;**33**(38):14989-14997. DOI: 10.1523/jneurosci.2039-13.2013

[27] Andrade MJO, Neto AC, Oliveira AR, Santana JB, Santos NA. Daily variation of visual sensitivity to luminance contrast: Effects of time of measurement and circadian typology. Chronobiology International. 2018b;**35**(7):996-1007. DOI: 10.1080/07420528.2018.1450753

[28] Bassi CJ, Powers MK. Daily fluctuations in the detectability of dim lights by humans. Physiology & Behavior. 1986;**38**(6):871-877. DOI: 10.1016/00319384(86)90056-9

[29] Tassi P, Pins D. Diurnal rhythmicity for visual sensitivity in humans? Chronobiology International. 1997;**14**:35-48. DOI: 10.3109/07420529709040540

[30] Tassi P, Pellerin N, Moessinger M, Hoeft A, Muzet A. Visual resolution

in humans fluctuates over the 24h period. Chronobiology International. 2000;**17**(2):187-195. DOI: 10.1081/CBI-100101042

[31] Andrade MJA, Cristino ED, Santos LGB, Oliveira AR, Santos AN. Daily variation of visual perception of colors: Preliminary studies. Psychology & Neuroscience. 2018a;**11**(3):238-251. DOI: 10.1037/pne0000132

[32] Foster RG, Kreitzman L. Rhythms of life: The biological clocks that control the daily lives of every living thing. New Haven, CT: Yale University Press; 2005

[33] Santos NA, Simas MLB. Contrast sensitivity function: Indicator of visual perception of shape and spatial resolution. Psychology: Reflection & Criticism. 2001;**14**(3):589-597. DOI: 10.1590/S0102-79722001000300014

[34] Turner PL, Mainster MA. Circadian photoreception: Ageing and the eye's important role in systemic health. Brazil Journal Ophthalmoly. 2008;**92**(11):1439-1444. DOI: 10.1136/bjo.2008.141747

[35] Hatori M, Panda S. The emerging roles of melanopsin in behavioral adaptation to light. Trends Molecular Medicine. 2010;**16**(10):435-446. DOI: 10.1016/j.molmed.2010.07.005

[36] Hattar S, Liao W, Takao M, Berson DM, Yau K. Melanopsin-containing retinal ganglion cells: architecture, projections, and intrinsic photosensitivity. Science. 2002;**295**:1065-1070. DOI: 10.1126/science.1069609

[37] O'Keefe LP, Baker HD. Diurnal changes in human psychophysical luminance sensitivity. Physiology Behavior. 1987;**41**:193-200

[38] Campbell FW, Maffei L. Contrast and spatial frequency. Scientific American. 1974;**231**:106-114

[39] De Valois RL, De Valois KK. Spatial Vision. New York: Oxford University Press; 1988

[40] Cornsweet TN. Visual Perception. New York: Academic Press; 1970

[41] Pelli DG, Tillman KA. The uncrowded window of object recognition. Nature Neuroscience. 2008;**11**(10):1129-1135. DOI: 10.1038/nn.2187

[42] Owsley C. Contrast sensitivity. Ophthalmology Clinics of North America. 2003;**16**:171-177. DOI: 10.1016/S0896-1549(03)00003-8

[43] Struck LK, Rodnitzky RL, Dobson JK. Circadian fluctuations of contrast sensitivity in Parkinson's disease. Neurology. 1990;**40**:467-470. DOI: 10.1212/WNL.40.3_Part_1.467

[44] Andrade MJO, Silva JA, Santo NA. Influence of chronotype and measure time on visual contrast sensitivity. Psicologia Reflexão & Crítica. 2015;**28**(3):522-531. DOI: 10.1590/1678-7153.201528311

[45] Blatter K, Cajochen C. Circadian rhythms in cognitive performance: Methodological constraints, protocols, theoretical underpinnings. Physiology & Behavior. 2007;**90**(2-3):196-208. DOI: 10.1016/j.physbeh.2006.09.009

[46] Hess RF, Hayes A. The coding of spatial position by the human visual system: effects of spatial scale and retinal eccentricity. Vision Research. 1994;**34**:625-643. DOI: 10.1016/0042-6989(94)90018-3

[47] Viola AU, Archer SN, James LM, Groeger JA, Lo JC, Skene DJ, et al. Per3

polymorphism predicts sleep structure and waking performance. Current Biology. 2007;**17**(7):613-618. DOI: 10.1016/j.cub.2007.01.073

[48] Carrasco M, McLean TL, Katz SM, Frieder KS. Feature asymmetries in visual search: Effects of display duration, target eccentricity, orientation & spatial frequency. Vision Research. 1998;**38**(3):347-374. DOI: 10.1016/S00426989(97)00152-1

[49] Mollon JD, Pokorny MJ, Knoblauch K. Normal and Defective Colour Vision. Oxford Scholarship: Oxford University Press; 2003

[50] Hankins MW, Lucas RJ. The primary visual pathway in humans is regulated according to long-term light exposure through the action of a nonclassical photopigment. Current Biology. 2002;**12**(3):191-198. DOI: 10.1016/S0960-9822(02)00659-0

[51] MacLeod DI. New dimensions in color perception. Trends in Cognitive Science. 2003;**7**:97-99. DOI: 10.1016/S1364-6613(03)00022-6

[52] Figueiro MG, Bullough JD, Parsons RH, Rea MS. Preliminary evidence for a change in spectral sensitivity of the circadian system at night. Journal Circadian Rhythms. 2005;**11**:3-14. DOI: 10.1186/1740-3391-3-14

[53] Lima MG, Gomes BD, Ventura DF, Silveira LCL. Methods used in psychophysical color evaluation. Psychology USP. 2011;**22**:197-222

[54] Neitz J, Neitz M. The genetics of normal and defective color vision. Vision Research. 2011;**51**(7):633-651. DOI: 10.1016/j.visres.2010.12.002

[55] Pauers MJ, Kuchenbecker JA, Neitz M, Neitz J. Changes in the colour of light cue circadian activity. Animal Behavior. 2012;**83**(5):1143-1151. DOI: 10.1016/j.anbehav.2012.01.035

[56] Danilenko KV, Plisov IL, Cooper HM, Wirz-Justice A, Hébert M. Human cone light sensitivity and melatonin rhythms following 24-hour continuous illumination. Chronobiology International. 2011;**28**(5):407-414. DOI: 10.3109/07420528.2011.567425

[57] Ebihara S, Tsuji K. Entrainment of the circadian activity rhythm to the light cycle: Effective light intensity for Zeitgeber in the retinal degenerate C3H mouse and the normal C57BL mouse. Physiology Behavior. 1980;**24**:523-527. DOI: 10.1016/00319384(80)90246-2

[58] Mrosovsky N. Contribution of classic photoreceptors to entrainment. Journal of Comparative Physiology. 2003;**189**(1):69-73. DOI: 10.1007/s00359-002-0378-7

[59] Hovis JK, Ramaswamy S. Color vision and fatigue: An incidental finding. Aviation, Space and Environmental Medicine. 2007;**78**(11):1068-1071. DOI: 10.3357/ASEM.2174.2007

[60] Archer SN, Robilliard DL, Skene DJ, Smits M, Williams A, Arendt J, et al. A length polymorphism in the circadian clock gene Per3 is linked to delayed sleep phase syndrome and extreme diurnal preference. Sleep: Journal of Sleep and Sleep Disorders Research. 2003;**26**(4):413-415. DOI: 10.1093/sleep/26.4.413

[61] Ventura DF, Silveira LCL, Rodrigues AR, De Souza JM, Gualtieri M, Bonci D, et al. Preliminary norms for the Cambridge colour test. In: Mollon JD, Pokorny J, Knoblauch K, editors. Normal and Defective Colour Vision. New York, NY: Oxford University

Press; 2003. pp. 536-547. DOI: 10.10939780198525301.003.0034

[62] Barrionuevo PA, Nicandro N, McAnany JJ, Zele AJ, Gamlin P, Cao D. Assessing rod, cone, and melanopsin contributions to human pupil flicker responses. Investigative Ophthalmology & Visual Science. 2014;**55**:719-727. DOI: 10.1167/iovs.13-13252

[63] Iuvone, P. M., Brown, A. D., Haque, R., Weller, J., Zawilska, J. B., Chaurasia, S. S., . . . Klein, D. C. (2002). Retinal melatonin production: Role of proteasomal proteolysis in circadian and photic control of arylalkylamine N-acetyltransferase. Investigative Ophthalmology & Visual Science, 43(2), 564-572.

[64] Walmsley L, Hanna L, Mouland J, Martial F, West A, Smedley AR, et al. Colour as a signal for entraining the mammalian circadian clock. PLoS Biology. 2015:1-20. DOI: 10.1371/journal.pbio.1002127

[65] Revell VL, Barrett DC, Schlangen LJ, Skene DJ. Predicting human nocturnal nonvisual responses to monochromatic and polychromatic light with a melanopsin photosensitivity function. Chronobiology International. 2010;**27**:1762-1777. DOI: 10.3109/07420528.2010.516048

Chapter 3

Circadian Modulation of Neurodevelopment in the Adult Human Brain: Importance of Melatonin

Héctor Solís-Chagoyán, Jairo Muñoz-Delgado, Rosa Estrada-Reyes, Salvador Alarcón-Elizalde and Gloria Benítez-King

Abstract

Melatonin (*N*-acetyl-5-methoxytryptamine) is an indoleamine synthesized by the pineal gland in the dark phase of the photoperiod. Released melatonin into the pineal recess and the cerebrospinal fluid is the chemical signal that conveys information about the environmental illumination to the brain. In recent years, it was described that melatonin stimulates the neurodevelopment in the adult brain. During this complex process, new neurons are formed and differentiate to form synaptic connections. Neuropsychiatric disorders are characterized by the loss of neuronal connectivity and diminished levels of melatonin, among other features. Importantly, these patients have impaired circadian rhythms. In recent years, evidence aroused indicating that neurodevelopment occurs in the adult brain, making important the study of chemical compounds and endogenous molecules that stimulate neurodevelopment to reestablish synaptic connectivity. In this chapter, we will review the evidence that supports the circadian melatonin modulatory effects on neurodevelopment and its importance for the treatment of neuropsychiatric diseases.

Keywords: circadian rhythms, melatonin, neurodevelopment, dendritogenesis, axogenesis, neuropsychiatric diseases

1. Introduction

The rotation of the Earth on its axis is the root of the 24-h light-dark cycle and all of the astronomical phenomena that are measured concerning the plane of the horizon—the sunrise and sunset, twilight periods, photoperiod, solar eclipses, movement of the tides, and the lunar perigee and apogee. These phenomena all have the common denominator that they cause variations in the periods of light versus darkness. This light-dark cycle and its variations directly influence the myriad of activities of living organisms, including sleep and wakefulness, rest and activity, feeding, and body temperature changes [1].

The timing and pattern of mammalian behavioral activities are regulated by an evolutionarily optimized interplay between the genetically based biological (circadian) clock and superimposed environmental factors and thus mask the effects mediated by the clock. The main regulator in endogenous circadian rhythms is the suprachiasmatic nucleus, which sits above the optic chiasm and, in humans, is composed of approximately 20,000 neurons. The most important external synchronizing factors, or "Zeitgebers," is the light-induced phase-setting of the circadian rhythmicity to the 24-h solar day. This influence results in a roughly 24-h activity-rest cycle in diurnal organisms and the converse rest-activity cycle in nocturnal organisms.

Although the 24-h light-dark cycle is the most important Zeitgeber, there are many other modulators that further synchronize or mask circadian rhythms, including social stimuli, sounds, smells, or physical contact, which generate behavioral states that impact endogenous clocks [2]. Regarding the light-dark cycle as a Zeitgeber, the production of melatonin (5-methoxitriptamine), an indolamine secreted by the pineal gland, is dependent on the photoperiod. This hormone is released at night by the pineal gland and collected by the internal recess, the third ventricle. From this location, melatonin is distributed by the cerebrospinal fluid to the brain tissue acting as a synchronizing signal of the internal media with the environmental light [3]. The circadian production of melatonin by the pineal gland is controlled by the circadian clock through a multi-synaptic pathway and released melatonin carries the timing information to the peripheral oscillating structures to couple the oscillating functions (**Figure 1**).

The role of melatonin as circadian messenger, as well as the organization of the circadian system, has been described mainly using confined animals; however, studying biological rhythms under natural conditions allows us to understand how geophysical variables impact endogenous clocks. In this regard, initially, chronobiologists conducted their studies under experimental laboratory conditions, while behavioral ecologists have focused their research on observing the behavior of free-living organisms. In the year 2000, in their book "Activity patterns in small mammals" Halle and Stenth of the University of Jena in Germany proposed the integration of behavioral ecology and chronobiology under natural and semi-natural conditions, founding the new discipline of "behavioral chrono ecology," that is, the study of rhythms under natural conditions and the synchronization of rhythms with nature [4].

In behavioral field studies of primates' activity rhythms and their modulation by environmental variables, the possible dual, synchronizing and/or masking effects of variables other than light-dark cycles are often ignored. However, there are some studies that addressed these issues. One of these was studied in the Mexican spider monkeys (*Ateles geoffroyi*) using long-term activity recordings with wearable accelerometer/data-logger devices (**Figure 2**). The relationship between astronomical and meteorological parameters and various parameters of the monkeys' rest-activity rhythm under semi-natural conditions was analyzed. The monkeys were exposed to natural light, temperature, and humidity cycles. By recording the monkeys' activity for 180 days, the monkeys rested during the night and were active throughout the day with two peaks in activity. Activity time, onset and end of activity, and the timing of their two activity peaks were significantly correlated with duration of the solar day and sunrise or sunset time. Beyond the light-dark cycle, weather, temperature, cloudiness, and artificial variables introduced by their interactions with humans also significantly influenced the duration and times of onset and end of activity [5].

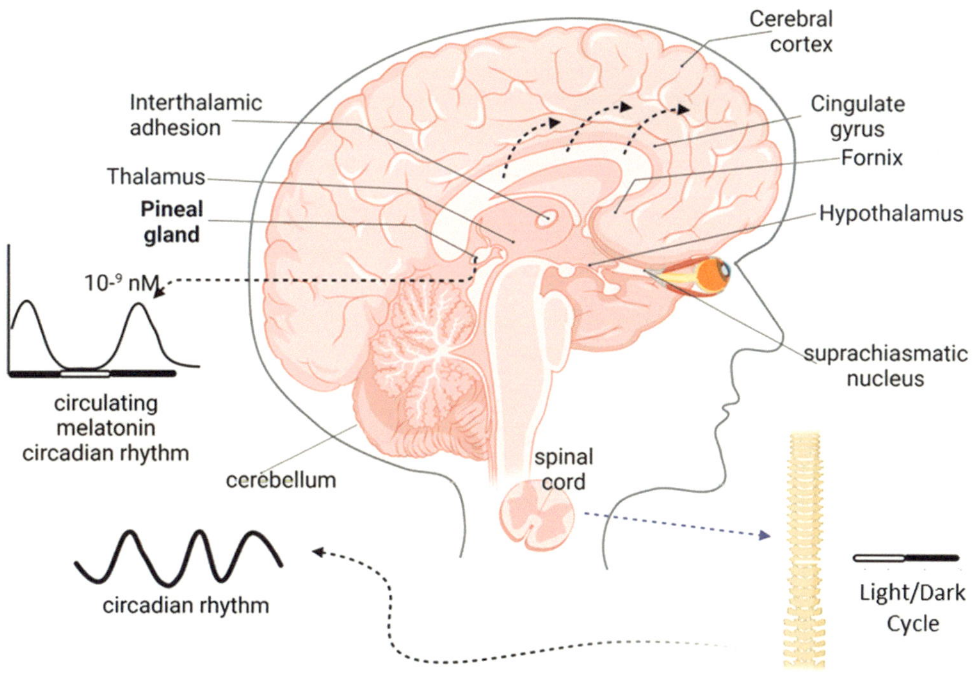

Figure 1.
Regulation pattern of melatonin synthesis. In the scheme is showed the principal structures involved in the circadian rhythm of melatonin synthesis, the suprachiasmatic nucleus and the pineal gland as well as the neuronal and the endocrine pathways to transmit the circadian information, the autonomous nervous system and the variation of circulating melatonin.

In other study, additional environmental variables such as the effect of housing conditions and season were incorporated to the daily timing and pattern of activity in this species. Thus, the activity patterns between monkeys living under natural lighting and climatic conditions in either a large wire netting cage or a 0.25-ha forest enclosure in the primatological field station of Veracruz State University near Catemaco, Mexico, were studied. Also, a pregnant female was followed in the forest enclosure, which gave us insight into the effect of late pregnancy and parturition on the monkey's activity rhythm. Spider moneys are strictly diurnal, with 90% of their total activity occurring during daylight. Monkeys that lived in the forest enclosure had a higher second activity peak in the late afternoon compared to those living in the caged area, resulting in a more pronounced bimodal activity pattern. The spider monkeys kept there had an earlier activity onset and morning activity peak than their conspecifics in the cage; however, no differences were found in the phase-setting parameters of the circadian system to the environmental 24-h periodicity, either by comparison or correlation with the sunrise and sunset. The late pregnancy, parturition, and lactation induced a reduction on the activity level during the week of parturition and the following week. The long days of the summer season and the short days of the winter season were decisive in the expression of the activity time of the morning and evening peaks. Together, data suggest that in Mexican spider monkeys, a weak circadian component and strong direct masking effects of multiple environmental factors are involved in the regulation of the daily activity rhythm [6].

The study of the monkeys under their natural environment may allow us to obtain evidence to highlight the importance of melatonin to modulate diverse oscillating

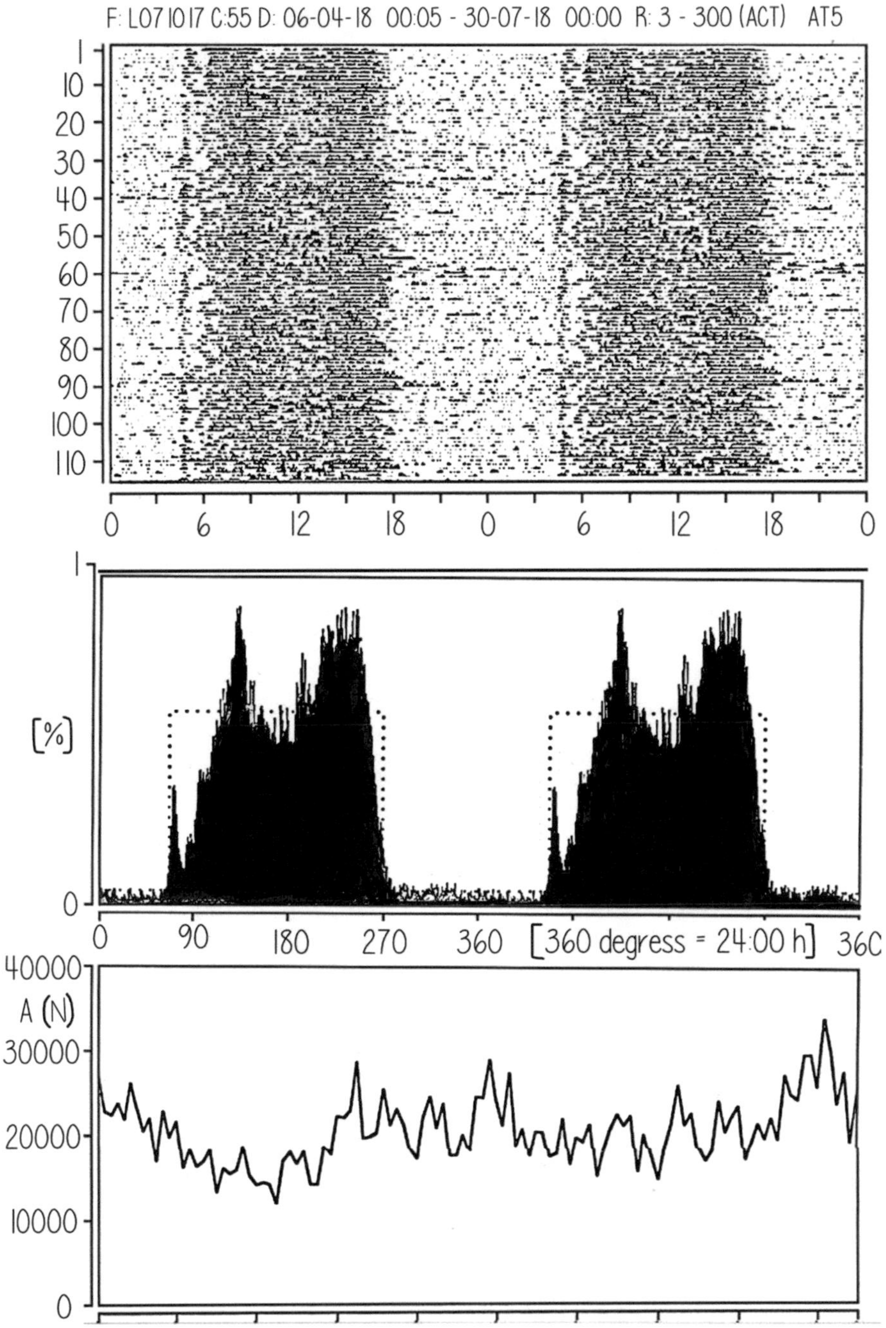

Figure 2.
Circadian rhythm of motor activity in the Mexican spider monkeys Ateles geoffroyi. The motor activity was recorded continuously and plotted as an actogram. This scheme shows the diurnal circadian pattern in this species.

functions and behavioral activities controlled by the circadian clock that can repercuss on mental health. Hence, the main goal of this chapter is to review and discuss the evidence that supports the influence of melatonin on circadian physiology focusing on its modulatory effects on neurodevelopment and brain plasticity as well as on the importance of this indole for the treatment of neuropsychiatric diseases.

2. Circadian rhythms can be modulated by melatonin

Melatonin (*N*-acetyl-5-methoxytryptamine) is a biogenic amine synthesized from tryptophan (**Figure 3**) that was first isolated and characterized by Aaron B. Lerner

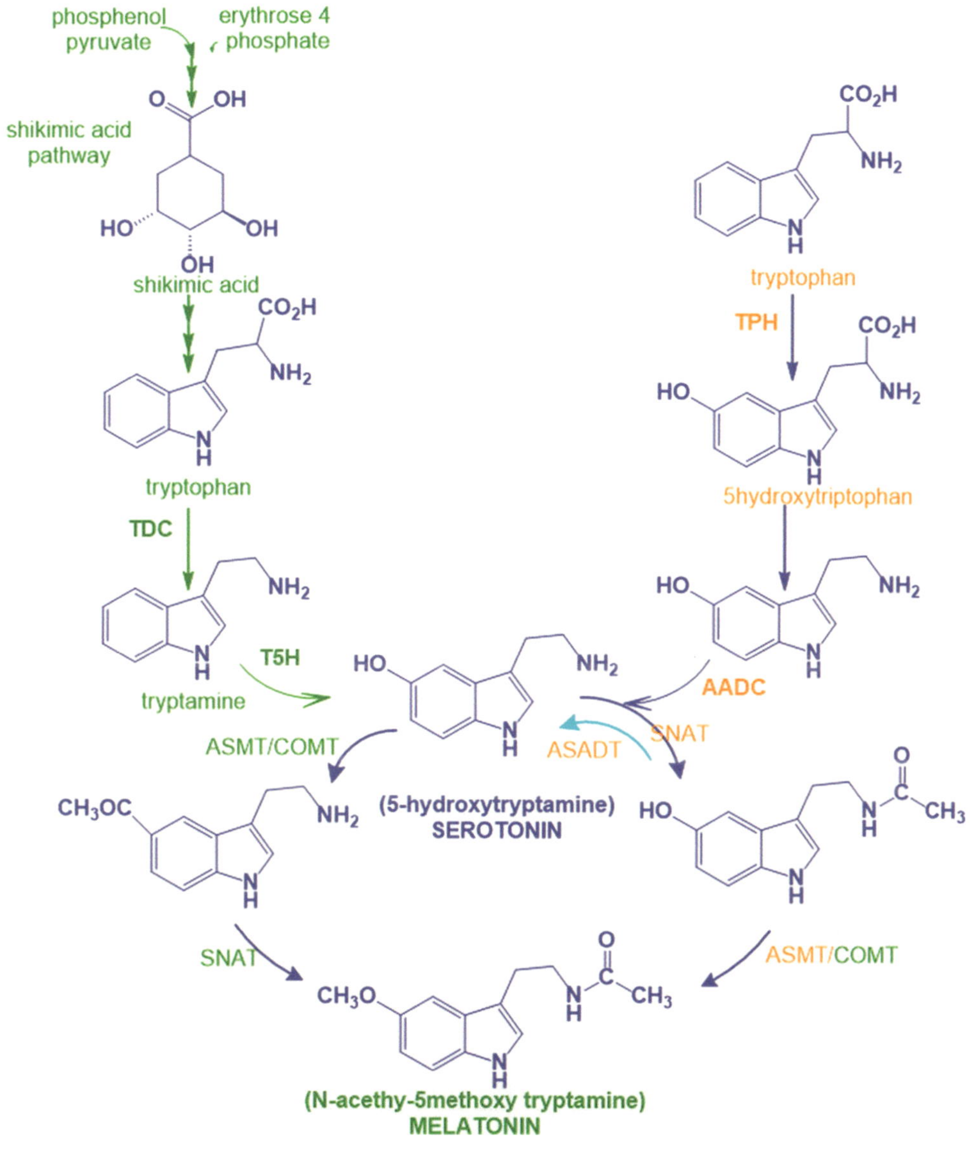

Figure 3.
Synthesis pathway of melatonin. The metabolic pathway for melatonin biosynthesis begins necessarily with tryptophan This synthesis consists of several enzymatic reactions; among them the hydroxylation by tryptophan hydroxylase (TPH) produces 5-hydroxytryptophan; in addition, serotonin is formed by the aromatic amino acid decarboxylase (AADC); at the end, N-acetylserotonin-O-methyltransferase (ASMT), produces melatonin.

and coworkers in 1958, from more than 200,000 lyophilized bovine pineal glands. The amphiphilic nature of melatonin allows it to cross the cytoplasmic membrane and interact directly with intracellular proteins as calmodulin [7, 8] or nuclear receptors [9]. This indole can also bind to membrane receptors to trigger specific signaling pathways [10]. This variety of molecular targets distributed inside cells as soluble proteins and outside cells such as heptahelical receptors could explain the extensive participation of melatonin to modulate a great variety of key functions in cellular physiology.

As mentioned previously, circulating melatonin concentration oscillates following a circadian rhythm with a nocturnal peak in almost all studied species (**Figure 4**). This oscillation is regulated by light implying that the rhythm could be synchronized by the photoperiod, in the same manner as other functions [11]. The most ancient function of melatonin apparently is the protection against oxidative stress. Still, in animals, this indole regulates even complex behaviors such as daily activity and seasonal reproduction, some sleep properties, as well as retinal, hormonal, metabolic, and immune functions [12]. In contrast to non-mammalian vertebrates, pinealectomy in rat or mouse does not disrupt the circadian rhythmicity; instead, these animals retain most of their circadian rhythms, including the oscillating locomotor activity; however, a subtle uncoupling of several physiological functions occurs [13]. In mammals, the melatonergic system has only a secondary rank in the circadian system [14]. Apparently, the temporal signal produced by SCN is distributed to peripheral clocks by both neural pathways through the hypothalamic nucleus and the autonomous nervous system [15, 16], and endocrine pathways mediated mainly through melatonin from the pineal gland. An example of the latter is the requirement of melatonin signal in the maintenance of circadian activity in the *pars tuberalis* of adenohypophysis [17].

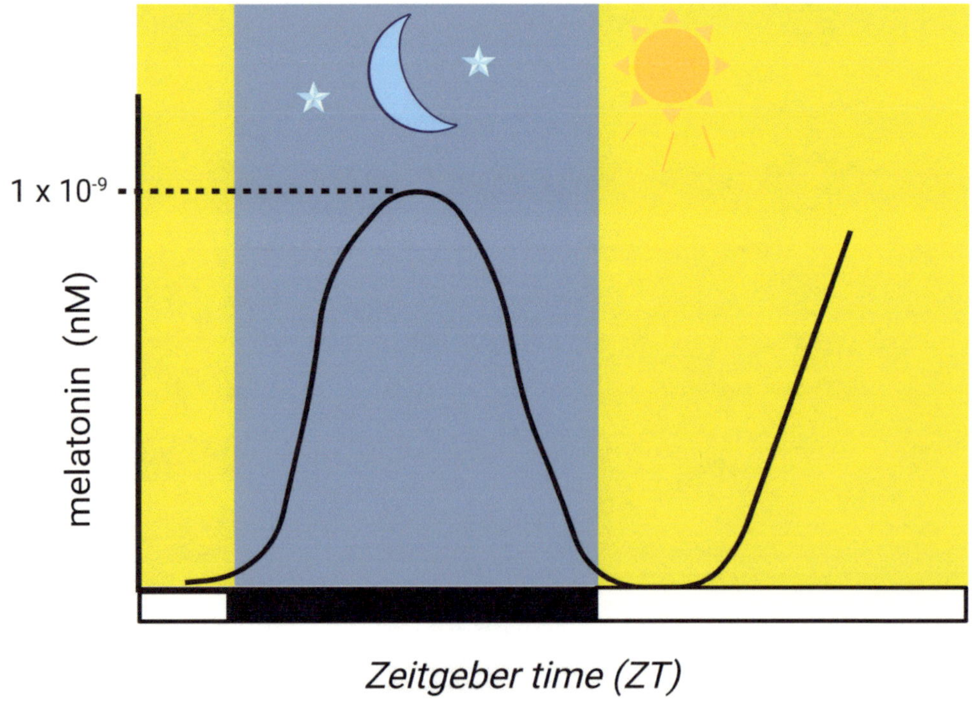

Figure 4.
Circadian rhythm of melatonin release. The hormone melatonin is released from the pineal gland in mammals at night. This molecule has been considered the signal of darkness to adjust the circadian system.

To address the role of melatonin as a synchronizer molecule, several experimental strategies have been developed. For example, the organisms are maintained under constant environmental conditions in which the circadian rhythms present a free-running period that could be shorter or longer than 24 h. In this condition, one stimulation with exogenous melatonin applied at different circadian times can induce a phase shift in the oscillation; this change can be a delay or an advance of the rhythm phase. Also, under the same constant environmental condition, a daily melatonin stimulus is applied at a specific external time, so the rhythm becomes adjusted with respect to the periodic stimulation. Both the period and the phase of the rhythms change and are synchronized by melatonin. In both cases, the conclusion is that melatonin is a synchronizer acting on the circadian pacemaker.

In all vertebrates, melatonin membrane receptors seem to be involved in both the phase shifts induced in the circadian oscillations and in the synchronization of the rhythms by injected or infused melatonin [18–20]. It has been proposed that the pathway activated by melatonin is dependent on the subtype of membrane receptors distributed in the SCN or in the pineal gland itself. In many non-mammalian vertebrates, melatonin receptors distributed in different tissues, including the SCN, seem to be involved in the transduction of the timing of circadian rhythms, but also in phase shifts and synchronization [20].

Melatonin shifts the phase of the locomotor activity rhythm and the firing rate of the mammalian SCN neurons. These actions can be mediated by MT2 receptors. The mammalian circadian pacemaker is sensitive to exogenous melatonin at the hours around day-night and night-day transitions [21, 22]. When melatonin is applied to male C3H/HeN mice at CT10, that is, two circadian hours before the onset of activity, the hormone induces dose-dependent phase advances [23]. This effect is significantly reduced by application of the selective MT2 antagonist 4P-PDOT. This antagonist also diminishes the melatonin-induced phase advances of the SCN neuronal firing rhythm in male Long-Evans rats [24]. In MT1 receptor-deficient mice, the application of melatonin at CT10 still produced phase advances [25]. A common example of circadian rhythms phase shift is the so-called jet lag induced by traveling across several time zones. Melatonergic agonists such as ramelteon and tasimelteon (nonselective MT1/MT2 agonists) have been used as therapeutic options for alleviating the sleep disturbances associated with this transient perturbation in the temporal organization of the circadian system [26].

Melatonin synchronizes the locomotor circadian rhythm in Long-Evans rats, Wistar rats, and the diurnal rodent *Arvicanthis ansorgei* following a daily stimulation procedure by injection or infusion of the indolamine [27–29]. Interestingly, in nocturnal rodents synchronized by melatonin, the activity onset is coincident with melatonin application, as expected considering that melatonin release is nocturnal; in contrast, in the synchronized diurnal species, melatonin application coincides with the beginning of the rest phase [30].

Even though pinealectomy does not induce apparent changes in the properties of most of the circadian rhythms in mammals, it has been observed that the decrease in melatonin levels affects some important circadian functions such as the sleep-wake cycle and triggers depressive-like symptoms; these disturbances have been improved with the administration of melatonin receptor agonists [31], suggesting that the role of this pleiotropic hormone in physiology might be underestimated. One important function of melatonin that can repercuss in the brain physiology is the modulation of neurogenic processes and circuit functioning as explained below.

3. The neuroplastic changes show circadian rhythms

Similarly, to the behavioral studies of primates' activity rhythms and their modulation by environmental variables, there are also studies about the regulation of molecular and cellular processes influenced by Zeitgebers that participate in the behavior. One example is the neurodevelopment in the hippocampus, which is the brain region where learning and memory are integrated [32]. This function plays a key role in the adaptation of the organism to the environment. Neurodevelopment implicates the formation of new neurons, cell migration, cell differentiation, and the formation of neuronal projections dendrites and axons as well as the formation of synaptic contacts that culminate in the establishment of neuronal connections that together constitute the structural network that underlies brain function.

In 1966, Altman found that new neuron formations occur in the adult brain, making a breakthrough in the concept of that times that new neuron formation occurred only at embryonic stages [33]. However, Altman, for the first time, demonstrated by immunohistology and autoradiographic technique the incorporation of 3H-thymidine into the DNA of cerebellar cells of rats at postnatal age [33]. In the last decade of the past century, further contributions to this field emerged, and it was clear that neurogenesis continues throughout the entire life at various locations in the brain such as the subventricular zone, the olfactory bulb, and the dentate gyrus of the hippocampus [34]. In an analogous manner to neurogenesis, dendrite and axonal formation as well as synaptic contact, establishment occurs during the entire life, making the brain a highly neuroplastic an adaptative organ.

Adult neurogenesis as well as neuroplasticity is conserved processes in all mammalian species studied so far including non-human primates and humans [35]. Both processes are mechanisms that allow the survival and the well adaptation of the organisms to environmental conditions. In this regard, melatonin, a phylogenetically conserved molecule, allows survival and adaptation acting as a free radical scavenger and enhancing the levels of antioxidant enzymes protecting the brain against stressful stimuli. Moreover, because the indolamine stimulates neurogenesis and new neuronal contact formation in the hippocampus, it makes the organism more competent for survival.

The evidence about circadian regulation of neurodevelopment in the adult brain was aroused by the observation that melatonin stimulates different stages of neurodevelopment and because the indolamine is synthesized according to the photoperiod and synchronized the internal activity with the environmental light. In this regard, melatonin synchronizes the sleep-wake cycle, the body temperature, the cortisol release, among other functions.

The adult brain presents neuroplastic changes that follows a circadian rhythm; for instance, the maximal amplitude of neuroplasticity occurs during the scotophase regardless of the activity pattern of the species (diurnal or nocturnal). In the hippocampal subgranular zone of murine adult, proliferation of immature neurons increases at the middle of the scotophase (*Zeitgeber Time* 18: *ZT*18) where the maximum amplitude of melatonin secretion occurs. By contrast, neuronal apoptosis decreases at that time having the maximal increase at the middle of the light time (*ZT*6). Interestingly, these rhythmic changes occur only in animals with functional melatonin MT1 and MT2 receptors [36].

In diurnal zebrafish, and in nocturnal mice, the stages of the cell cycle in stem cells of neurogenic niches show a circadian rhythm with nocturnal peak [37]. These rhythms correlate with the expression of Clock1 and Per1 in zebrafish [37] and with

Per2 and Bmal1 in mice [38]. These results suggest that cell proliferation is controlled by the clock genes in most mammalian species [39–41]; interestingly, melatonin can influence the level of clock gene expression [42] and the half-life of clock proteins by timing their degradation in the proteasome [43].

Neuroplastic changes in dendrites also follow a circadian pattern. In Siberian hamsters, the hippocampal dendritic structure of basilar dendrites of CA1pyramidal neurons revealed dendrite length increase that occurs during the dark phase [44]. Moreover, the number of branch points significantly increases during short days (8 h L: 16 h D), indicating a more complex dendritic arbor, while during long days (16 h L: 8 h D), the dendrites are longer. Administration of melatonin at the end of the light phase induces the nocturnal dendritic morphology in CA1 neurons within 4 h. In addition, in organotypic cultures of hippocampus incubated with melatonin 100 nM, dendrite formation in the hilar zone is evident after 6 h of incubation and increased formation of secondary and tertiary dendrites [45] (**Figure 5**). The changes in dendrite size and complexity are correlated with the increase in the expression of Per1 and Bmal1 in the hippocampus [46]. Hence, in short or long photoperiod schemes, the duration of endogenous melatonin release would change respecting the length of the dark phase, providing seasonal information to neurogenic niches [47].

Synaptogenesis, the main event by which neurons are connected, is also regulated by the circadian rhythms. Two approaches have been used to study synaptic formation: (1) structural studies in which immunostaining of synaptic proteins by specific antibodies was used to label dendrite spines the site where synaptogenesis takes place and electrophysiological approaches where long-term potentiation is studied (LTP). The former showed a circadian pattern reaching a maximum number or size of dendritic spines during the dark phase in the hippocampus and somatosensory cortex in mice [48, 49]. Moreover, increased synapses formation is observed in hippocampal organotypic cultures

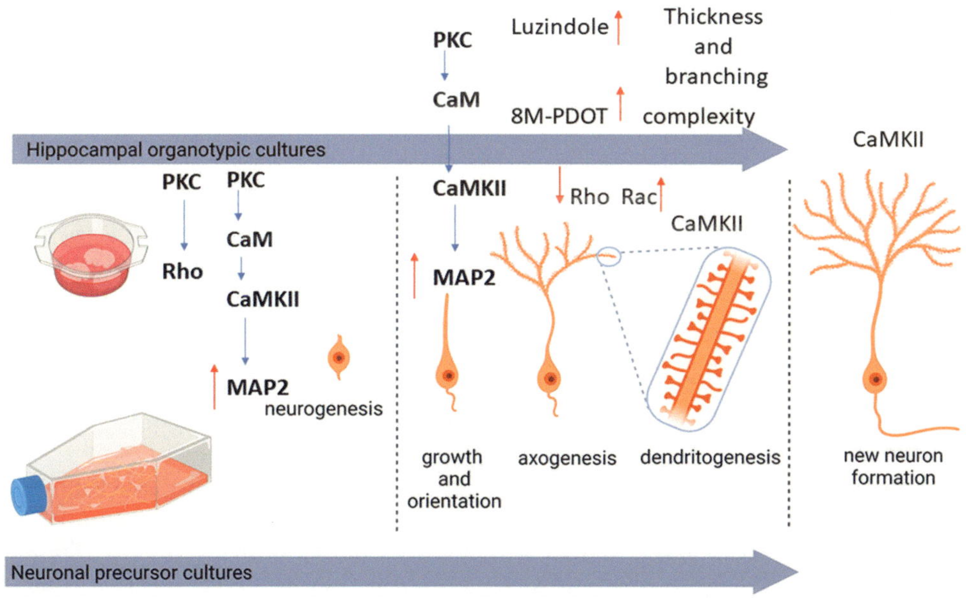

Figure 5.
Neurogenesis in hippocampus and neuronal precursor cells. The neurogenesis process has been studied in organotypic cultures and in isolated precursors obtained from the olfactory epithelium to determine the stimulatory effect of melatonin.

incubated with 100 nM melatonin and by staining with an anti-synapsin antibody, which labels synapsin a protein localized in the presynapsis [45]. Despite this information, studies in Siberian hamsters indicate a decrease in dendritic spine density during the dark phase in the dentate gyrus [44, 46]. Thus, it is necessary to study more deeply the rhythmicity of synapses formation and the factors that influence it.

On the other hand, electrophysiological studies have shown that synaptic strength measured through hippocampal LTP was more significant in magnitude and stability in slices obtained during the dark phase [50]. This pattern persisted even in constant darkness or in slices obtained at daytime and recording LTP at night; these results suggest an endogenous rhythm in synaptic plasticity that could persist *ex vivo* in the slices. Moreover, melatonin, perfused during the light phase, reduces the hippocampal LTP at a wide range of concentrations (0.01–100 μM) [51, 52]. In summary, these results suggest that hippocampal neuroplasticity is controlled by the light/dark cycle and by melatonin. Other evidence concerning melatonin's role as neuronal modulator has been found in neuronal stem cells obtained from the olfactory epithelium, as mentioned in the next section.

4. Melatonin stimulatory effects on structural neuroplasticity in a translational model

Current studies indicate that olfactory neuronal precursors are a subrogate model of the central nervous system to study neurotransmitter receptors expression, enzymes involved in neurotransmitter synthesis as well as the neurodevelopment in the adulthood.

Olfactory neurons have a common ectodermal embryonic origin with the CNS neurons and are derived from embryonic placodes and the neural crest, which are structures analogous to the neural tube [53, 54].

Gene expression profiles of olfactory neuronal precursors are similar to mesenchymal stem cells and can be differentiated into mature olfactory neurons and other types of neurons such as dopaminergic neurons [55, 56]. The expression of hundreds of genes such as CNS neurons has been shown in olfactory neuroepithelial cells. Some of these are the pituitary adenylate cyclase-activating peptide and the glutamate receptor, among others [57–59]. Moreover, in postmortem samples obtained from Alzheimer's disease patients, paired helical filaments of tau protein and amyloid-β plaques similar to those found in cortical and subcortical neurons have been described in olfactory neuroepithelial cells characterized by cytokeratin-18 expression reflecting its stromal epithelial cell nature, as well as in olfactory neurons characterized by III β-tubulin expression [60, 61]. Recently, neuronal precursors were obtained from alive patients with Alzheimer's disease diagnosis and demonstrated the paired helical filaments, as well as increased levels of tau total and phospho-tau supporting that olfactory neuroepithelial cells are a mirror model that reflects molecular changes produced in the CNS and useful to study different stages of neurodevelopment in samples obtained from alive patients [62].

In this regard, it was demonstrated that spontaneous axogenesis occurs in olfactory neuronal precursors derived from a clone obtained by unlimited dilution from a female of 55 years with no psychiatric history. Twelve percent of these cells maintained in primary culture the ability to form axons. In the presence of melatonin, axonal formation augmented by 15% in the primary cultures of olfactory neuronal precursors [63].

In addition to these observations, spine and new neuron formation in a clone of human olfactory neuronal precursors stimulated by melatonin was demonstrated. Spines were labeled with Phalloidin-Rhodamine and an anti-spinophilin antibody and counted. The prelaminar results showed that melatonin increases spine formation in primary cultures of human olfactory neuronal precursors. Moreover, neurogenesis measured as clusters of proliferating neuronal precursors has been observed in the presence of 100 nM melatonin (manuscript in preparation). Altogether, evidence supports that melatonin stimulates three important stages of neurodevelopment, neurogenesis, spine formation, and axogenesis.

5. Chrono-disruption in neuropsychiatric diseases

In addition to natural *Zeitgebers* and masking factors, certain pathologies and the pharmaceuticals used to treat them can also alter activity patterns. It is well known that disruption of circadian cycles is a common feature in neuropsychiatric diseases. Among these, alterations in the sleep-wake cycle, core body temperature, appetite, and hormonal release cycles such as the cortisol secretion have been described in patients with neuropsychiatric alterations such as Alzheimer disease, bipolar disorder, major depression, and schizophrenia among other diseases. Importantly, all these functions are cyclically modulated by melatonin and in neuropsychiatric patients, the amplitude of melatonin secretion peak is blunted.

In human subjects, the pattern of motor activity in unmedicated schizophrenia patients and healthy subjects had been study to disclose whether the pattern was affected by treatment with typical and atypical antipsychotics (haloperidol and risperidone). Twenty unmedicated schizophrenic patients wore a wrist actigraph that recorded activity at 1-min intervals for five days. The activity pattern of patients and healthy subjects was compared; then, the patients were randomly assigned to treatment with low-dose haloperidol or risperidone, and the effect of treatment on the activity was tested. Untreated patients were less active during morning, evening, and late-night periods compared with controls. Both haloperidol and risperidone induce significant effects on activity level (this effect was more prominent with haloperidol). The results suggest that unmedicated schizophrenic patients exhibit abnormally low levels of motor activity, which persists with antipsychotic treatment, even though symptoms improve. Future studies should clarify whether motor disturbances are a primary effect of the illness or related to the lifestyle effects [64].

In addition, some antidepressants flatten the amplitude of melatonin secretion, contributing to the alterations of biological rhythms inherent to major depression. Thus, it can be suggested the use of melatonin as a therapeutical adjuvant to regulate biological rhythms such as the sleep/wake cycle.

Recently, we explored if neuroplasticity is altered in schizophrenia. By using the translational experimental model of olfactory neuronal precursors, we found that these cells derived from a patient with schizophrenia were unable to form axons spontaneously. However, in the presence of melatonin, these cells formed axons with the same rate as cells derived from a healthy subject with no psychiatric history [63]. Thus, data suggest that besides the inherent disruption of biological rhythms in schizophrenia patients and drug administration, they also have impaired axonal rhythm formation during adulthood that can be restored by melatonin.

6. Importance of time administration of drugs in the treatment of neuropsychiatric diseases

As mentioned before, human beings have a rhythmic expression in their physiology and behavior. In homeostasis states, this endogenous rhythmicity is synchronized by cyclic environmental factors. This coupling provides several advantages: anticipating cyclical changes in the background and facilitating the organism's adaptation to its environment. This rhythmicity has systems that allow measuring the passage of time, and it is regulated by environmental signals (exogenous) that act as external synchronizers. These rhythms oscillate approximately every 24 h, and among these, the most important is the light-dark cycle or cycle circadian. This concerted, internal, and external rhythmic expression is essential to maintain a healthy state. There is strong evidence that supports the idea that the disruption or chrono-disruption of this adaptive mechanism is detrimental to health and has, among other consequences, sleep disorders, which can lead to cognitive deterioration. It has also been associated with an increased risk of cardiovascular disease, hypertension, metabolic conditions such as diabetes, cancer, obesity, and affective disorders such as anxiety and depression, increasing the prevalence of neuropsychiatric disorders such as schizophrenia and major depression.

The synchronization of circadian rhythms with pharmacotherapy is crucial for neuropsychiatric and affective disorders treatment since it is about combining the maximum benefit with the minimum time complexity of the treatments.

Adequate coordination between drug administration schedules to obtain an optimal therapeutic response has been little explored and currently represents a challenge for chrono-pharmacology.

Desynchronization between the times of administration and the potential pharmacological effects could be one of the reasons why a high percentage of patients with major and bipolar depression (MD) are resistant to treatment, as well as the chronic recurrence of depressive and seasonal disorders. Chronotherapeutic strategies that reset the internal clock may have a specific advantage for treating depression and other mental disorders.

For instance, seasonal affective disorder (SAD) is a sub-type of depression in which individuals experience depressive symptoms and show hypersomnia only in the winter months, in which the period of darkness is more extended than at other times of the year. In a pioneering study, Rosenthal and coworkers [65] found that treatment with bright environmental light suppresses the endogen melatonin secretion and reverses the winter depressive symptoms of patients with SAD. In contrast, light too dim to suppress endogen melatonin is therapeutically ineffective.

This same paper described the antidepressant effects of phototherapy in eight SAD patients by oral melatonin administration. However, in another study with 19 SAD patients, the authors did not find any therapeutic difference between the atenolol, a beta-adrenergic blocker, which inhibits melatonin secretion, and placebo. In contrast, research with seven SAD patients showed that the antidepressant effects of phototherapy were not photoperiodic and appeared to be independent of melatonin suppression. Authors conclude that melatonin can mediate the effects of shortening days on the winter symptoms of SAD and that the modification of melatonin secretion by bright light mediates its antidepressant effects and gives evidence that melatonin secretion may be abnormal in SAD [65].

Multiple factors can disrupt this chronicity because of imbalances in the sleep-wake cycle. Disruption of circadian cycles has also been associated with affective disorders.

Depression disorders are characterized by a broad range of symptoms, including altered mood, loss of cognitive functions, and recurrent thoughts of death or suicide. The relationship between chrono-disruption and the etiology of depression is not yet clear. However, evidence suggests that existing pharmacotherapies such as lithium and antidepressants such as melatonin and agomelatine act on the circadian system.

Although it is known that melatonin is a mediator of photoperiodic changes on seasonal rhythms in animals, a gradual increase in circulating levels of melatonin occurs after lights off, reaching its maximum around the middle of the dark phase. There is contradictory evidence about the antidepressant effect of melatonin itself. We found that the melatonin administration in mice at two *Zeitgeber times* (*ZT* = 0 lights on; 12:12 L/D), 1 h before the beginning (*ZT*11) and at the middle (*ZT*18) of the dark phase after either a single or a three-dose treatment, produces a robust antidepressant-like effect in the tail suspension and the forced swimming tests. When a single dose of melatonin (4 mg/kg) was administered at *ZT* 11 produced an antidepressant-like effect in two paradigms. However, when melatonin was administered at *ZT* 18, it was ineffective in the forced swimming test. Required a higher dose of melatonin (16 mg/kg) to observe their antidepressant effect in the tail suspension test. In contrast, repeated doses of melatonin (*ZT* 18, *ZT* 11, and *ZT* 18) were necessary to produce the antidepressant effect in the forced swimming test. These results highlight the importance of the timely administration of melatonin could improve its antidepressant-like effect [66].

Agomelatine is an antidepressant that acts as an agonist on the melatonin receptors and as a 5HT2C receptor antagonist. Several studies suggest that the antidepressant effect caused by agomelatine results from the resynchronization of circadian rhythms that are disturbed in depressed patients ameliorating symptoms of depression. In contrast to other antidepressants, this has shown low relapse rates upon discontinuation and high tolerability. Furthermore, agomelatine treatment improves the amplitude of the circadian (rest/activity) sleep/wake cycle and diminishes the depression and anxiety symptoms in comparison with sertraline treatment [67].

There is strong evidence about the association between sleep disturbances and depression; in depressive disorder (MDD), the desynchronization of circadian rhythms occurs, producing disturbed sleep and insomnia, and these symptoms improve markedly with melatonergic (MT1 and MT2) and 5HT2C agonist treatment such as melatonin and agomelatine, which act as modulating the circadian rhythmicity.

The foregoing gives evidence of the disruptions of the sleep-wake cycle (sleep architecture and timing) and residual symptoms may prevent the attainment of high-quality remission and delay recovery from MDD.

Benedetti and coworkers studied the effect of morning light therapy or placebo combined with the serotonin reuptake inhibitor citalopram in treating patients affected by a major depressive episode without psychotic features. They found that the combination of this antidepressant and light treatment was more effective than citalopram alone or placebo in the treatment of major depression, administered with an optimized timing of administration, and low-intensity light treatment that significantly hastened and potentiated the effect of citalopram. This evidence provides the clinical psychiatrists with an augmenting strategy, effective and devoid of side effects [68].

Alterations in circadian rhythmicity have been little studied in patients with schizophrenia, in which sleep disorders are common, with an 80% prevalence that often responds to circadian disruption.

This study showed that the variability of sleep-wake time is notably more significant and more remarkable in the schizophrenia group than in the people without the disorder [69].

In addition, polysomnography studies have shown that these patients present a higher sleep latency, diminished total sleep time, lower efficiency, more interruptions, a shorter duration, and latency of REM dreams, as well as a lower proportion of slow-wave sleep than people without this condition. Also, deficit sleep spindles in schizophrenia people could be an endophenotype of this disorder.

In a study of rest activity, a cohort of patients with schizophrenia matched with healthy subjects, Wulff and coworkers [69] showed that there is clear evidence of sleep and circadian rhythm disruption in schizophrenia patients, over half the cohort, tested showed severe circadian misalignment, including the melatonin cycle.

Compared rest-activity patterns in a cohort of patients with schizophrenia with matched healthy unemployed controls showed significant sleep/circadian disruption in all 20 patients studied. Of these, half showed severe circadian misalignment in sleep-wake and melatonin cycles, demonstrating that abnormal entrainment of the circadian system is prevalent in schizophrenia.

Although the results are not conclusive yet and more research is needed in this regard, chronotherapy is a relatively recent proposal to optimize pharmacological treatments based on the biological clock to maximize the pharmacological response or produce adjustments when there is a desynchronization in the biological functions that are affected in illnesses, including mental disorders. Therefore, it is necessary to adequately coordinate medication administration schedules to obtain the maximum pharmacological results and increase treatment adherence.

7. Conclusion

Evidence presented in this chapter indicates that neuroplasticity is modulated by an external *Zeitgeber*, the photoperiod, and synchronized by melatonin which is an endogenous synthesized molecule during the dark phase. Notably, functions such as neurogenesis, dendrite, axonal, and synaptogenesis are upregulated during the dark phase of the photoperiod and by 100 nanomolar concentrations of melatonin, which is the concentration reached in the cerebrospinal fluid during the dark phase of the photoperiod. Nowadays, neuroplasticity is considered an important tool for the treatment of neuropsychiatric diseases to repair damaged circuitry in the brain of these subjects. Therefore, since melatonin can increase all the stages of neuroplasticity at pharmacological concentrations, it is possible for the use of this indolamine as an adjuvant for the treatment of neuropsychiatric diseases where chrono-disruption is one of the main symptoms evidenced by alterations in the sleep/wake cycle, the motor activity and the pineal secretion of melatonin among other rhythms important for the good quality of life.

Acknowledgements

Funded by Consejo Nacional de Ciencia y Tecnología (CONACyT) Grant No. 290526 to GBK. CONACyT had no further role in review design, in the collection, analysis and interpretation of data, neither in the writing of the report, nor in the decision to submit the paper for publication.

Conflict of interest

The authors declare no conflict of interest.

Author details

Héctor Solís-Chagoyán[1], Jairo Muñoz-Delgado[2], Rosa Estrada-Reyes[3], Salvador Alarcón-Elizalde[1] and Gloria Benítez-King[1*]

1 Laboratory of Neuropharmacology, Subdirection of Clinical Research, National Institute of Psychiatry Ramón de la Fuente Muñiz, CDMX, Mexico

2 Laboratory of Chronoecology and Human Ethology, Department of Ethology, Direction of Neurosciences, National Institute of Psychiatry Ramón de la Fuente Muñiz, CDMX, Mexico

3 Laboratory of Phytopharmacology, Direction of Neurosciences, National Institute of Psychiatry Ramón de la Fuente Muñiz, CDMX, Mexico

*Address all correspondence to: bekin@imp.edu.mx

References

[1] Muñoz-Delgado J. A propósito de los ritmos de la Luna. Ciencia. Revista de la Academia Mexicana de Ciencias. 2008;**59**:48-56

[2] Mistlberger RE, Skene DJ. Social influences on mammalian circadian rhythms: Animal and human studies. Biological Reviews of the Cambridge Philosophical Society. 2004;**79**:533-556

[3] Hardeland R. Recent findings in melatonin research and their relevance to the CNS. Central Nervous System Agents in Medicinal Chemistry. 2018;**18**:102-114

[4] Halle S, Stenseth NC. Activity Patterns in Small Mammals. Berlin: Springer-Verlag; 2000. p. 322

[5] Muñoz-Delgado J, Corsi-Cabrera M, Canales-Espinosa D, Santillán-Doherty AM, Erkert HG. Astronomical and meteorological parameters and rest-activity rhythm in the spider monkey *Ateles geoffroyi*. Physiology & Behavior. 2004;**83**:107-117

[6] Muñoz-Delgado J, Sánchez-Ferrer JC, Pérez-Galicia S, Canales-Espinosa D, Erkert HG. Effects of housing conditions and season on the activity rhythm of spider monkeys (*Ateles geoffroyi*) kept under natural conditions within their distributional range in Central Mexico. Chronobiology International. 2014;**31**:983-995

[7] Antón-Tay F, Martínez I, Tovar R, Benítez-King G. Modulation of the subcellular distribution of calmodulin by melatonin in MDCK cells. Journal of Pineal Research. 1998;**24**:35-42

[8] Benítez-King G, Huerto-Delgadillo L, Antón-Tay F. Binding of 3H-melatonin to calmodulin. Life Sciences. 1993; **53**:201-207

[9] Smirnov AN. Nuclear melatonin receptors. Biochemistry (Mosc). 2001;**66**:19-26

[10] Sallinen P, Saarela S, Ilves M, Vakkuri O, Leppäluoto J. The expression of MT1 and MT2 melatonin receptor mRNA in several rat tissues. Life Sciences. 2005;**76**:1123-1134

[11] Pévet P, Bothorel B, Slotten H, Saboureau M. The chronobiotic properties of melatonin. Cell and Tissue Research. 2002;**309**:183-191

[12] Zhao D, Yu Y, Shen Y, Liu Q, Zhao Z, Sharma R, et al. Melatonin synthesis and function: Evolutionary history in animals and plants. Frontiers in Endocrinology (Lausanne). 2019;**10**:249

[13] Pévet P, Agez L, Bothorel B, Saboureau M, Gauer F, Laurent V, et al. Melatonin in the multi-oscillatory mammalian circadian world. Chronobiology International. 2006;**23**:39-51

[14] Stehle JH, von Gall C, Korf HW. Melatonin: A clock-output, a clock-input. Journal of Neuroendocrinology. 2003;**15**:383-389

[15] Buijs RM, Kalsbeek A. Hypothalamic integration of central and peripheral clocks. Nature Reviews. Neuroscience. 2001;**2**:521-526

[16] Kalsbeek A, Buijs RM. Output pathways of the mammalian suprachiasmatic nucleus: Coding circadian time by transmitter selection and specific targeting. Cell and Tissue Research. 2002;**309**:109-118

[17] von Gall C, Weaver DR, Moek J, Jilg A, Stehle JH, Korf HW. Melatonin

plays a crucial role in the regulation of rhythmic clock gene expression in the mouse pars tuberalis. Annals of the New York Academy of Sciences. 2005;**1040**:508-511

[18] Foà A, Magnone C, Bertolucci C. Circadian organization in ruin lizards: Phase response curve for melatonin changes with season. Journal of Comparative Physiology. A, Neuroethology, Sensory, Neural, and Behavioral Physiology. 2002;**188**:141-145

[19] Hyde LL, Underwood H. Daily melatonin infusions entrain the locomotor activity of pinealectomized lizards. Physiology & Behavior. 1995;**58**:943-951

[20] Pfeffer M, Korf HW, Wicht H. Synchronizing effects of melatonin on diurnal and circadian rhythms. General and Comparative Endocrinology. 2018;**258**:215-221

[21] Benloucif S, Masana MI, Dubocovich ML. Responsiveness to melatonin and its receptor expression in the aging circadian clock of mice. The American Journal of Physiology. 1997;**273**:R1855-R1860

[22] McArthur AJ, Hunt AE, Gillette MU. Melatonin action and signal transduction in the rat suprachiasmatic circadian clock: Activation of protein kinase C at dusk and dawn. Endocrinology. 1997;**138**:627-634

[23] Dubocovich ML, Yun K, Al-Ghoul WM, Benloucif S, Masana MI. Selective MT2 melatonin receptor antagonists block melatonin-mediated phase advances of circadian rhythms. The FASEB Journal. 1998;**12**:1211-1220

[24] Hunt AE, Al-Ghoul WM, Gillette MU, Dubocovich ML. Activation of MT(2) melatonin receptors in rat suprachiasmatic nucleus phase advances the circadian clock. American Journal of Physiology. Cell Physiology. 2001;**280**:C110-C118

[25] Liu C, Weaver DR, Jin X, Shearman LP, Pieschl RL, Gribkoff VK, et al. Molecular dissection of two distinct actions of melatonin on the suprachiasmatic circadian clock. Neuron. 1997;**19**:91-102

[26] Srinivasan V, Singh J, Pandi-Perumal SR, Brown GM, Spence DW, Cardinali DP. Jet lag, circadian rhythm sleep disturbances, and depression: The role of melatonin and its analogs. Advances in Therapy. 2010;**27**:796-813

[27] Slotten HA, Krekling S, Pévet P. Photic and nonphotic effects on the circadian activity rhythm in the diurnal rodent *Arvicanthis ansorgei*. Behavioural Brain Research. 2005;**165**:91-97

[28] Slotten HA, Krekling S, Sicard B, Pévet P. Daily infusion of melatonin entrains circadian activity rhythms in the diurnal rodent *Arvicanthis ansorgei*. Behavioural Brain Research. 2002;**133**:11-19

[29] Redman J, Armstrong S, Ng KT. Free-running activity rhythms in the rat: Entrainment by melatonin. Science. 1983;**219**:1089-1091

[30] Pevet P, Challet E. Melatonin: Both master clock output and internal time-giver in the circadian clocks network. Journal of Physiology, Paris. 2011;**105**:170-182

[31] Mairesse J, Silletti V, Laloux C, Zuena AR, Giovine A, Consolazione M, et al. Chronic agomelatine treatment corrects the abnormalities in the circadian rhythm of motor activity and sleep/wake cycle induced by

prenatal restraint stress in adult rats. The International Journal of Neuropsychopharmacology. 2013; **16**:323-338

[32] Lisman J, Buzsáki G, Eichenbaum H, Nadel L, Ranganath C, Redish AD. Viewpoints: How the hippocampus contributes to memory, navigation and cognition. Nature Neuroscience. 2017;**20**:1434-1447

[33] Altman J, Das GD. Autoradiographic and histological studies of postnatal neurogenesis. I. A longitudinal investigation of the kinetics, migration and transformation of cells incorporating tritiated thymidine in neonate rats, with special reference to postnatal neurogenesis in some brain regions. The Journal of Comparative Neurology. 1966;**126**:337-389

[34] Lim DA, Alvarez-Buylla A: The adult ventricular-subventricular zone (V-SVZ) and olfactory bulb (OB) neurogenesis. Cold Spring Harbor Perspectives in Biology. 2016;**8**:a018820

[35] Knoth R, Singec I, Ditter M, Pantazis G, Capetian P, Meyer RP, et al. Murine features of neurogenesis in the human hippocampus across the lifespan from 0 to 100 years. PLoS One. 2010;**5**:e8809

[36] Fredrich M, Hampel M, Seidel K, Christ E, Korf HW. Impact of melatonin receptor-signaling on Zeitgeber time-dependent changes in cell proliferation and apoptosis in the adult murine hippocampus. Hippocampus. 2017;**27**:495-506

[37] Akle V, Stankiewicz AJ, Kharchenko V, Yu L, Kharchenko PV, Zhdanova IV. Circadian kinetics of cell cycle progression in adult neurogenic niches of a diurnal vertebrate. The Journal of Neuroscience. 2017; **37**:1900-1909

[38] Bouchard-Cannon P, Mendoza-Viveros L, Yuen A, Kærn M, Cheng HY. The circadian molecular clock regulates adult hippocampal neurogenesis by controlling the timing of cell-cycle entry and exit. Cell Reports. 2013;**5**: 961-973

[39] Bieler J, Cannavo R, Gustafson K, Gobet C, Gatfield D, Naef F. Robust synchronization of coupled circadian and cell cycle oscillators in single mammalian cells. Molecular Systems Biology. 2014;**10**:739

[40] Janich P, Meng QJ, Benitah SA. Circadian control of tissue homeostasis and adult stem cells. Current Opinion in Cell Biology. 2014;**31**:8-15

[41] Feillet C, van der Horst GT, Levi F, Rand DA, Delaunay F. Coupling between the circadian clock and cell cycle oscillators: Implication for healthy cells and malignant growth. Frontiers in Neurology. 2015;**6**:96

[42] Imbesi M, Arslan AD, Yildiz S, Sharma R, Gavin D, Tun N, et al. The melatonin receptor MT1 is required for the differential regulatory actions of melatonin on neuronal 'clock' gene expression in striatal neurons in vitro. Journal of Pineal Research. 2009;**46**:87-94

[43] Vriend J, Reiter RJ. Melatonin feedback on clock genes: A theory involving the proteasome. Journal of Pineal Research. 2015;**58**:1-11

[44] Ikeno T, Weil ZM, Nelson RJ. Photoperiod affects the diurnal rhythm of hippocampal neuronal morphology of *Siberian hamsters*. Chronobiology International. 2013;**30**:1089-1100

[45] Domínguez-Alonso A, Ramírez-Rodríguez G, Benítez-King G. Melatonin increases dendritogenesis in

the hilus of hippocampal organotypic cultures. Journal of Pineal Research. 2012;**52**:427-436

[46] Ikeno T, Nelson RJ. Acute melatonin treatment alters dendritic morphology and circadian clock gene expression in the hippocampus of *Siberian hamsters*. Hippocampus. 2015;**25**:142-148

[47] Reiter RJ. The melatonin rhythm: Both a clock and a calendar. Experientia. 1993;**49**:654-664

[48] Ikeda M, Hojo Y, Komatsuzaki Y, Okamoto M, Kato A, Takeda T, et al. Hippocampal spine changes across the sleep-wake cycle: Corticosterone and kinases. The Journal of Endocrinology. 2015;**226**:M13-M27

[49] Jasinska M, Jasek-Gajda E, Woznicka O, Lis GJ, Pyza E, Litwin JA. Circadian clock regulates the shape and content of dendritic spines in mouse barrel cortex. PLoS One. 2019;**14**:e0225394

[50] Chaudhury D, Wang LM, Colwell CS. Circadian regulation of hippocampal long-term potentiation. Journal of Biological Rhythms. 2005;**20**:225-236

[51] El-Sherif Y, Tesoriero J, Hogan MV, Wieraszko A. Melatonin regulates neuronal plasticity in the hippocampus. Journal of Neuroscience Research. 2003;**72**:454-460

[52] Wang LM, Suthana NA, Chaudhury D, Weaver DR, Colwell CS. Melatonin inhibits hippocampal long-term potentiation. The European Journal of Neuroscience. 2005;**22**:2231-2237

[53] Forni PE, Wray S. Neural crest and olfactory system: New prospective. Molecular Neurobiology. 2012;**46**:349-360

[54] Sidman RL, Rakic P. Neuronal migration, with special reference to developing human brain: A review. Brain Research. 1973;**62**:1-35

[55] Delorme B, Nivet E, Gaillard J, Häupl T, Ringe J, Devèze A, et al. The human nose harbors a niche of olfactory ectomesenchymal stem cells displaying neurogenic and osteogenic properties. Stem Cells and Development. 2010;**19**:853-866

[56] Horiuchi Y, Kano S, Ishizuka K, Cascella NG, Ishii S, Talbot CC Jr, et al. Olfactory cells via nasal biopsy reflect the developing brain in gene expression profiles: Utility and limitation of the surrogate tissues in research for brain disorders. Neuroscience Research. 2013;**77**:247-250

[57] Hegg CC, Lucero MT. Dopamine reduces odor- and elevated-K(+)-induced calcium responses in mouse olfactory receptor neurons in situ. Journal of Neurophysiology. 2004;**91**:1492-1499

[58] Kanekar S, Gandham M, Lucero MT. PACAP protects against TNFα-induced cell death in olfactory epithelium and olfactory placodal cell lines. Molecular and Cellular Neurosciences. 2010;**45**:345-354

[59] Thukral V, Chikaraishi D, Hunter DD, Wang JK. Expression of non-N-methyl-D-aspartate glutamate receptor subunits in the olfactory epithelium. Neuroscience. 1997;**79**:411-424

[60] Arnold SE, Lee EB, Moberg PJ, Stutzbach L, Kazi H, Han LY, et al. Olfactory epithelium amyloid-beta and paired helical filament-tau pathology in Alzheimer disease. Annals of Neurology. 2010;**67**:462-469

[61] Ayala-Grosso CA, Pieruzzini R, Diaz-Solano D, Wittig O, Abrante L,

Vargas L, et al. Amyloid-aβ peptide in olfactory mucosa and mesenchymal stromal cells of mild cognitive impairment and Alzheimer's disease patients. Brain Pathology. 2015;**25**:136-145

[62] Riquelme A, Valdés-Tovar M, Ugalde O, Maya-Ampudia V, Fernández M, Mendoza-Durán L, et al. Potential use of exfoliated and cultured olfactory neuronal precursors for in vivo Alzheimer's disease diagnosis: A Pilot Study. Cellular and Molecular Neurobiology. 2020;**40**:87-98

[63] Galván-Arrieta T, Trueta C, Cercós MG, Valdés-Tovar M, Alarcón S, Oikawa J, et al. The role of melatonin in the neurodevelopmental etiology of schizophrenia: A study in human olfactory neuronal precursors. Journal of Pineal Research. 2017;**63**:1-16

[64] Apiquian R, Fresán A, Muñoz-Delgado J, Kiang M, Ulloa RE, Kapur S. Variations of rest-activity rhythm and sleep-wake in schizophrenic patients versus healthy subjects: An actigraphic comparative study. Biological Rhythm Research. 2008;**39**:69-78

[65] Rosenthal NE, Sack DA, Jacobsen FM, James SP, Parry BL, Arendt J, et al. Melatonin in seasonal affective disorder and phototherapy. Journal of Neural Transmission. Supplementum. 1986;**21**:257-267

[66] Estrada-Reyes R, Valdés-Tovar M, Arrieta-Baez D, Dorantes-Barrón AM, Quero-Chávez D, Solís-Chagoyán H, et al. The timing of melatonin administration is crucial for its antidepressant-like effect in mice. International Journal of Molecular Sciences. 2018;**19**:2278

[67] Kasper S, Hajak G, Wulff K, Hoogendijk WJ, Montejo AL, Smeraldi E, et al. Efficacy of the novel antidepressant agomelatine on the circadian rest-activity cycle and depressive and anxiety symptoms in patients with major depressive disorder: A randomized, double-blind comparison with sertraline. The Journal of Clinical Psychiatry. 2010;**71**:109-120

[68] Benedetti F, Colombo C, Pontiggia A, Bernasconi A, Florita M, Smeraldi E. Morning light treatment hastens the antidepressant effect of citalopram: A placebo-controlled trial. The Journal of Clinical Psychiatry. 2003;**64**:648-653

[69] Wulff K, Dijk DJ, Middleton B, Foster RG, Joyce EM. Sleep and circadian rhythm disruption in schizophrenia. The British Journal of Psychiatry. 2012;**200**:308-316

Chapter 4

Circadian Synchrony between Mothers and Young in the European Rabbit: Or Not? A Cautionary Tale

Robyn Hudson and Gerard A. Kennedy

Abstract

The European rabbit *Oryctolagus cuniculus*, ancestor of all domestic breeds, has an unusual pattern of maternal care in which females briefly nurse their young just once approximately every 24 h, and where the pups anticipate and prepare for their mother's arrival. Chronobiologists have seen this as a model mammalian system to study the physiological and neurobiological underpinnings of a biologically relevant circadian complex. However, observations of nursing in wild rabbits, together with studies of nursing in domestic breeds allowed free access to their young in laboratory settings, suggest that the rabbit's pattern of daily nursing visits resembles an hourglass rather than a circadian process, well suited to the sudden starts and stops of natural nursing cycles. We consider whether there might be other such cases in the literature, including in human chronobiology, in which failing to consider the organism's natural, evolved daily patterns of behaviour and prematurely studying these under artificially imposed laboratory time schedules might have also led to such patterns being erroneously considered circadian.

Keywords: daily rhythm, circadian, hourglass, maternal behaviour, mother-young synchrony, ontogeny, rabbit, *Oryctolagus cuniculus*

1. Introduction

"Oh dear! Oh dear! I shall be too late". The White Rabbit in Alice's Adventures in Wonderland, Lewis Carroll (1865) [1].

1.1 The rabbit as a model species in mammalian chronobiology

Timing is clearly of the essence in many aspects of the European rabbit's behavioural biology and contributes importantly to this species' proverbial reproductive success. The European rabbit *Oryctolagus cuniculus*, the ancestor of all domestic breeds, has been widely used in the study of mammalian chronobiology. Consistent with various well-studied rodent models, such as the white rat (*Rattus norvegicus*) and the golden hamster (*Mesocricetus auratus*), the rabbit shows circadian (or at least diurnal) rhythmicity in several important behavioural and physiological functions. These include motor activity, feeding, drinking, urination, and defecation, as well as haematological parameters, serotonin concentration in the brainstem, content and absorption of volatile fatty acids in the alimentary tract, visual evoked potentials, and intraocular pressure (review in [2]). In addition, due to the rabbit's abundance, size, and importance as an agricultural pest, it is one of the best-studied laboratory mammals in the wild.

Under natural conditions, the rabbit is a primarily nocturnal or crepuscular species, which is most active during the night with peaks in activity levels at dawn and dusk (in nature: [3–5]; in the laboratory: [6, 7]). Aschoff's rule states that the endogenous free-running circadian period observed in complete darkness (DD) will shorten for diurnal animals and lengthen for nocturnal animals when they are exposed to constant light [8]. Rabbits typically show free-running rhythms in several functions that lengthen in accordance with Aschoff's rule when they are transferred from DD to LL [9]. Furthermore, the phase-response curve of the rabbit's daily pattern of wheel-running activity in response to brief 1-hour light pulses [10] conforms to classic phase-response curves reported for nocturnal rodents [11].

2. The rabbit's unusually limited pattern of maternal care: once-daily nursing

Interest in the rabbit as a mammalian model in circadian studies has been reinforced by this species' unusual pattern of maternal care. In the wild, rabbits give birth to litters comprising several altricial young in an underground nursery burrow that the mother digs either within or close to the communal "warren" [12, 13]. Shortly before giving birth, she constructs a nest in this burrow (or laboratory nest box) of dried grass and fur pulled from her chest and flanks (review in [14]). Parturition usually lasts only a few minutes [15, 16], after which the mother immediately leaves the young, closes and disguises the burrow entrance, and only returns to quickly re-open it and nurse the pups for a few minutes approximately once every 24 hours. The end of nursing is signalled by the mother jumping away from the pups, the pups dropping immediately from the nipples [16, 17], and the mother immediately leaving the burrow (or nest box) and closing the entrance after each visit until the approach of weaning [12, 13].

During nursing, the mother stands over the pups without giving them any direct assistance to locate the nipples and suckle (see 2.1). If the mother is pregnant with a subsequent litter, which is often the case due to post-partum oestrus and her mating again immediately after giving birth, she will abruptly stop nursing at around post-partum day 26 and wean her young in preparation for the birth of the next litter several days later ([18, 19], review in [20]). This pattern of brief once-daily nursing visits is shown both by wild rabbits and by domestic breeds in conditions of husbandry and the laboratory (wild rabbit: [13, 21, 22]; domestic rabbit: [23–26]; review in [27]), and has generally been interpreted as a strategy to reduce the risk of the rabbit's many predators locating the open burrow and trapping the nursing mother and young there, and allowing the mother to forage more widely for food [18, 24, 27].

It is a pattern that can also be readily replicated in the laboratory by separating mothers from their pups and only allowing them access to the nest to nurse for a few minutes once each day. Mothers accept this regimen well and raise their young without apparent difficulty (but see Section 2.5). Furthermore, if mothers are given a second opportunity to nurse several hours after the first nursing of the day, they fail to do so [24, 25].

DOI: http://dx.doi.org/10.5772/intechopen.101922

2.1 Synchrony between mothers and young

Such limited maternal care, so different to the extensive care characteristic of most mammalian mothers, is only possible due to several adaptations on the part of the young, synchronising their behaviour with that of their mother. The pups anticipate their mother's nursing visit 1 or 2 hours before her arrival with increased motor activity, resulting in them uncovering from the nest material and thereby gaining unimpeded access to her nipples [25, 28]. They also show an anticipatory rise in body temperature [29] and changes in several endocrine and metabolic parameters [30–33].

Mothers also show a pre-nursing rise in body temperature [34]. In addition, they emit chemical cues from their ventrum, a so-called nipple-search pheromone, which elicits an inborn, stereotyped and highly effective pattern of searching for, attaching to and sucking nipples by the pups immediately the mother positions herself over them ([17, 35]; review in [36]). While nursing, the mother remains motionless, during which she shows a large release of oxytocin into the bloodstream, stimulating a single large milk-ejection reflex, and enabling the pups to drink up to a quarter of their body mass in less than the 3 to 4 minutes, during which she remains positioned over them [17, 37, 38]. At the end of each of the first few visits, she also deposits several hard faecal pellets in the nest, which the pups start to consume after several days, although she never urinates there [19]. Following nursing, the pups simultaneously urinate, become wet, and vigorously dig back under the nest material, fluffing it up and in the process becoming dry again [25].

2.2 A model of circadian rhythmicity including during early development

This unusual pattern of behaviour in such an important aspect of mammalian reproductive biology has attracted the interest of chronobiologists both because of the practical advantages it offers for experimentation, and also as it provides such a clear example of a daily rhythm with a biologically relevant, adaptive function. Interest was soon accompanied by classical chronobiological experiments from various laboratories, seeming to confirm the circadian nature of the mother's nursing rhythm and associated functions, and also of the pups' anticipatory arousal ([25, 28, 39]; reviews in [2, 14, 27]). As mentioned in Section 2.1, mothers allowed access to their pups to nurse at the same time each day show an anticipatory rise in body temperature, and the pups also, while the pups also show an anticipatory increase in motor activity and in uncovering from the nest material. Additionally, in the case of the pups, when the anticipated nursing visit is omitted (possible as pups readily survive missing one nursing, and at later ages even two nursings), they soon return to their baseline behavioural and physiological levels, and then approximately 24 h later (approximately 47 h after the last nursing) again show the usual anticipatory patterns, suggesting these are under the control of endogenous circadian processes ([25, 28, 29, 34]; review in [2]).

Excitement at the seeming evidence for circadian rhythmicity in the relation between mothers and young in the experimentally amenable rabbit quickly resulted in a series of demanding and sophisticated studies of the neural and molecular processes underlying such rhythms, that is, of the regulation of such endogenous biological "clocks", both in mothers and young. These studies, however, have so far provided somewhat equivocal results.

2.3 Mothers

In the case of the mothers, several studies have been undertaken to locate the nuclei of the neural circuits thought to represent the substrate for the mother's nursing rhythm, although without finding strong support for this being regulated by the rabbit's circadian system. Findings have included both the lack of nursing-induced expression of the "clock" gene protein PER1 [40, 41] or enhanced c-Fos expression in the suprachiasmatic nucleus of the hypothalamus [42], generally considered to be the master clock regulating circadian rhythms in mammals [43]. One reason for these equivocal findings might be that the rabbit's nursing rhythm, despite early assumptions, is not part of the rabbit's circadian system and is regulated by processes largely independent of this (see Section 2.5).

2.4 Young

Despite the value of a developmental approach to understanding biological systems, there have been relatively few studies of the development of circadian phenomena in neonatal mammals (but see [44–46]). One reason for this is the extensive maternal care shown by most mammals that makes it difficult to exclude the contribution of the mother, or other caregivers, to patterns of activity observed in the young. Partly for this reason, the behaviour of rabbit pups was early proposed as a model to study the ontogeny of the rabbit's circadian system [28, 39, 47]. Again, various molecular biological techniques such as the expression of c-Fos and various "clock" genes in the pup hypothalamus have been used in the attempt to identify the structures in the developing brain potentially regulating the anticipatory behaviour and physiological functions associated with the once-daily nursing, and the persistence of such patterns in the absence of the mother's nursing visit [48–51]. Interesting though the findings of these studies are, their significance for an understanding of the development of the rabbit's circadian system is unclear. This is because as outlined below (Section 2.5), the pups' anticipation of and preparation for their mother's daily visit seems to be the product of an hourglass mechanism, reset at each nursing, rather than of an endogenous circadian process.

2.5 Truly circadian or rather an hourglass process?

Despite the circadian enthusiasm, reports started to emerge quite early suggesting that perhaps the female rabbit's daily pattern of nursing and the accompanying daily pattern of pups' anticipatory arousal were not, in fact, a circadian-regulated package. One cause for doubt were observations in wild rabbits that mothers did not return to their pups on a 24-hour basis, but rather during the days following parturition returned with a periodicity shorter than 24 hours, arriving a little earlier at each visit. As mothers typically give birth early in the day, this resulted in their nursing visits drifting back into the night ([13, 21, 22]; review in [27]). A second cause for doubt was the timing of nursing by mothers in the laboratory allowed free access to their young. They showed the same pattern as wild rabbits, giving birth in the daylight hours and arriving to nurse a little earlier on each visit, and so with their visits also drifting back into the night-time ([27]; cf. [52]). This gradual shift apparently does not disrupt the synchrony between mothers and their young, since the pups anticipate their mother's arrival by an hour or so, allowing them to be prepared for her earlier arrival each time. Following nursing, they then apparently reset their "clock" a little earlier in anticipation of her next visit.

Additional experiments then suggested why this gradual separation between the timing of birth and the timing of nursing might be physiologically important to mothers, and that nursing visits might be timed so as not to interfere with physiological processes associated with pregnancy and parturition. Specifically, there is evidence [52] that this might be to avoid the large surge in the release of oxytocin into the mother's bloodstream during nursing ([37]; see also ([18]) from provoking premature parturition in pregnant mothers at a time of day when the (pregnant) uterus is maximally sensitive to oxytocin [53], and when the swift parturition characteristic of the rabbit [16] is made possible by a single large release of oxytocin into her bloodstream [15].

The above findings are consistent with the results of a further series of experiments in which rabbit mothers maintained under controlled laboratory conditions were allowed scheduled or free access to their young, and their nursing visits compared to the circadian pattern of their feeding behaviour. When allowed free nest box access, mothers' nursing visits also had a periodicity shorter than 24 hours, cutting across, and apparently independent from the circadian regularity of their daily pattern of feeding [54]. Furthermore, when the pups were permanently removed from the nest before the normal age of weaning and the box left open, mothers stopped visiting the nest within 1 or 2 days, showing little evidence of an endogenous, self-sustaining circadian nursing rhythm.

Together, these findings suggest that rather than the mother rabbit's nursing rhythm (and by implication the associated behaviour of her pups) being regulated by a circadian mechanism, it corresponds to an hourglass process more appropriate to, more adaptive for "stop and go" functions such as nursing and weaning than to the inertia of a self-sustaining rhythm. For certain functions, an hourglass mechanism, allowing a quick response to rapidly changing short-term contingencies may be more adaptive than an enduring, self-sustaining circadian mechanism. In the rabbit, such contingencies include the abrupt cessation of nursing visits in late lactation by mothers pregnant with a further litter [18, 19], or in response to nest mortality, in which mothers may lose an entire litter, for example, due to predation, infanticide or flooding [55–57]. Under such circumstances it would be presumably maladaptive for mothers to repeatedly return to raided or flooded nests, and when they should return to breeding as soon as possible.

2.6 What you get out is what you put in. Circularity?

To the extent that the above is correct, it suggests that supposed circadian rhythmicity in the rabbit's daily pattern of nursing and associated patterns of behaviour in the young might be an artefact of protocols that have applied experimental procedures giving the animals little option other than to confirm the experimenters' assumptions of the operation of circadian mechanisms in the regulation of the behaviour and physiology of mothers and young in the nursing context. Thus, when given no choice, mothers show anticipatory behaviour as reported previously, enter the nest box (sometimes somewhat frantically, e.g. [54]) to nurse their young every 24 hours, and the young anticipate this as previously reported. However, this line of evidence involves a certain circularity; mothers (and again by extension their young) are typically given no choice other than to confirm the assumption of circadian regulation upon which experimental protocols of limited, 24-hour access to the young have been based. When allowed free access to their young, whether under natural conditions in the field or the laboratory, mothers arrive with a period somewhat less than 24 hours and show a nursing rhythm that drifts back, cutting across their well-established, approximately 24-hour rhythm in feeding activity [54]. This is further confirmed by Apel et al. [54] in nursing mothers exposed to LD cycles at the limits of their range of entrainment [10], where nursing visits show a periodicity of around 23 hours irrespective of the LD cycle period. Furthermore, when the young pups are permanently removed from the nest, thereby removing feedback to the mother from stimuli associated with suckling such as stimulation of the nipples ([58, 59]; see also [19, 60, 61]), nursing visits abruptly cease within approximately 24 hours. As suggested in Section 2.4, the regular diurnal nursing visits of rabbit mothers thus correspond better to the operation of an hourglass than to a circadian process ([54]; see review in [62]).

3. Conclusions

The caution implied in our title refers to the danger of not considering a study species' natural behaviour and the ecological conditions under which it evolved when designing experimental procedures under often highly artificial laboratory conditions and in interpreting the results obtained from these within highly reductionist frameworks. In the present case of the rabbit's unusual daily pattern of nursing, this has arguably involved jumping too quickly to assume that this and associated patterns of behaviour and physiological and neural functions, including in the young, to have a circadian basis, and then investing considerable time and money to explore underlying physiological, neural and molecular mechanisms based on these possibly erroneous assumptions. At times, it has also involved an important confusion in the use of terminology between circadian and diurnal processes in article titles and abstracts, and thus in the interpretation of results. If our above historical analysis of the case of the rabbit's nursing rhythm is correct, it raises the question that in how many other models of "circadian" function, including in humans, might such bias also be the case? In mammalian studies, and in contrast to studies in invertebrates, the adaptive advantages of flexible hourglass processes, set to start and stop by immediate environmental events, have been little considered in contrast to the benefits but also costs of the inertia, the lag in adjustment, of circadian processes (review in [62]). Speculatively, in the human case, one can think of the ease (and pleasure) with which many of us transition from workday routines to often very different weekend or holiday schedules. And to complicate things, since we are dealing with the regulation of biological functions in complex organisms leading complex lives, perhaps we need to consider that diurnal functions may be regulated by a combination of circadian and hourglass processes working in tandem?

Author details

Robyn Hudson[1*] and Gerard A. Kennedy[2]

1 National University of Mexico, Mexico City, Mexico

2 Federation University, Ballarat, Australia

*Address all correspondence to: rhudson@iibiomedicas.unam.mx

References

[1] Carroll L. Alice's Adventures in Wonderland. London: Macmillan; 1865

[2] Jilge B, Hudson R. Diversity and development of circadian rhythms in the European rabbit. Chronobiology International. 2001;**18**:1-26. DOI: 10.1081/CBI-100001275

[3] Mykytowycz R, Rowley I. Continuous observations of the activity of the wild rabbit, *Oryctolagus cuniculus* (L.) during 24-hour periods. CSIRO Wildlife Research. 1958;**3**:26-31. DOI: 10.1071/CWR9580026

[4] Kolb HH. Circadian activity in the wild rabbit (*Oryctolagus cuniculus*). Mammal Review. 1986;**16**:145-150. DOI: 10.1111/j.1365-2907.1986.tb00035.x

[5] Wallage-Drees JM. A field study on seasonal changes in the circadian activity of rabbits. Zeitschrift für Säugetierkunde. 1989;**54**:22-30

[6] Pivik RT, Bylsma FW, Cooper P. Sleep-wakefulness rhythms in the rabbit. Behavioral & Neural Biology. 1986;**45**:275-286. DOI: 10.1016/SO163-1047(86)80016-4

[7] Jilge B. The rabbit: A diurnal or a nocturnal animal? Journal of Experimental Animal Science. 1991; **34**:170-183

[8] Aschoff J. Exogenous and endogenous components in circadian rhythms. Cold Spring Harbor Symposium Quarterly Review of Biology. 1960;**25**:11-28. DOI: 10.1101/sqb.1960.0025.01.004

[9] Jilge B, Stähle H. The internal synchronization of five circadian functions of the rabbit. Chronobiology International. 1984;**1**:195-204. DOI: 10.3109/07420528409063896

[10] Kennedy GA, Hudson R. Phase response curve to 1 h light pulses for the European rabbit (*Oryctolagus cuniculus*). Chronobiology International. 2016;**33**:1120-1128. DOI: 10.1080/07420528.2016.1191506

[11] Daan S, Pittendrigh CS. A functional analysis of circadian pacemakers in nocturnal rodents. II. The variability of phase response curves. Journal of Comparative Physiology A. 1976;**106**: 253-266. DOI: 10.1007/BF01417857

[12] Lloyd HG, McCowan D. Some observations on the breeding burrows of the wild rabbit, *Oryctolagus cuniculus*, on the island of Skokholm. Journal of Zoology. 1968;**156**:540-549. DOI: 10.111/j.1469-7998.1968.tb04376.x

[13] Rödel HG, Dausmann KH, Starkloff A, Schubert M, von Holst D, Hudson R. Diurnal nursing pattern of wild-type European rabbits under natural breeding conditions. Mammalian Biology. 2012;**77**:441-446. DOI: 10.1016/j.mambio.2012.04.002

[14] González-Mariscal G, Caba M, Martínez-Gómez M, Bautista A, Hudson R. Mothers and offspring: The rabbit as a model system in the study of mammalian maternal behavior and sibling interactions. Hormones and Behavior. 2016;**77**:30-41. DOI: 10.1016/j.yhbeh.2015.05.011

[15] Fuchs AR, Dawood MY. Oxytocin release and uterine activation during parturition in rabbits. Endocrinology. 1980;**107**:1117-11126. DOI: 10.1210/endo-107-4-1117

[16] Hudson R, Cruz Y, Lucio RA, Ninomiya J, Martínez-Gómez M. Temporal and behavioral patterning of parturition in rabbits and rats.

Physiology & Behavior. 1999;**66**:599-604. DOI: 10.1016/S0031-9384(98)00331-X

[17] Hudson R, Distel H. Nipple location by newborn rabbits: Behavioural evidence for pheromonal guidance. Behaviour. 1983;**85**:260-275. DOI: 10.1163/156853983X00255

[18] Lincoln DW. Suckling: A time-constant in the nursing behaviour of the rabbit. Physiology & Behavior. 1974;**13**:711-714. DOI: 10.1016/0031-9384(74)90247-9

[19] Hudson R, Bikó Á, Altbäcker V. Nursing, weaning and the development of independent feeding in the rabbit (*Oryctolagus cuniculus*). Zeitschrift für Säugetierkunde. 1996;**61**:39-48. Available from: https://www.biodiversitylibrary.org/permissions/

[20] Martínez-Gómez M, Juárez M, Distel H, Hudson R. Overlapping litters and reproductive performance in the domestic rabbit. Physiology & Behavior. 2004;**82**:629-636. DOI: 10.1016/j.physbeh.2004.05.011

[21] Broekhuizen S, Mulder JL. Differences and similarities in nursing behaviour of hares and rabbits. Acta Zoologica Fennica. 1983;**174**:61-63

[22] Broekhuizen S, Bouman E, Went W. Variations in time of nursing in the brown hare (*Lepus europaeus*) and the European rabbit (*Oryctolagus cuniculus*). Mammal Review. 1986; **16**:139-144. DOI: 10.1111/j.1365-2907.1986.tb00034.x

[23] Deutsch JA. Nest building behaviour of domestic rabbits under semi-natural conditions. British Journal of Animal Behaviour. 1957;**5**:53-54. DOI: 10.1016/S0950-5601(57)80026-4

[24] Zarrow MX, Dennenberg VH, Anderson CO. Rabbit: Frequency of suckling in the pup. Science. 1965;**150**:1835-1836. DOI: 10.1126/science.150.3705.1835

[25] Hudson R, Distel H. The pattern of behaviour of rabbit pups in the nest. Behaviour. 1982;**79**:255-271. DOI: 10.1016/15685392X00292

[26] Coureaud G, Schaal B, Coudert P, Hudson R, Rideaud P, Orgeur P. Mimicking natural nursing conditions promotes early pup survival in domestic rabbits. Ethology. 2000;**106**:207-225. DOI: 10.1046/j.1439-0310.2000.00521.x

[27] Hudson R, Distel H. The temporal pattern of suckling in rabbit pups: A model of circadian synchrony between mother and young. In: Reppert SM, editor. Development of Circadian Rhythmicity and Photoperiodism in Mammals. Boston (USA): Perinatology Press; 1989. pp. 83-102

[28] Jilge B. The ontogeny of circadian rhythms in the rabbit. Journal of Biological Rhythms. 1993;**8**:247-260. DOI: 10.1177/074873049300800307

[29] Jilge B, Kuhnt B, Landerer W, Rest S. Circadian thermoregulation in suckling rabbit pups. Journal of Biological Rhythms. 2000;**15**:329. DOI: 10.1177/074487300012900l431

[30] Escobar C, Hudson R, Martínez-Gómez M, Aguilar-Roblero R. Metabolic correlates of the circadian pattern of suckling-associated arousal in young rabbits. Journal of Comparative Physiology A. 2000;**186**:33-38. DOI: 10.10071s003590050004

[31] Rovirosa MJ, Levine S, Gordon MK, Caba M. Circadian rhythm of corticosterone secretion in the neonatal rabbit. Developmental Brain Research. 2005;**158**:92-96. DOI: 10.1016/j.devbrainres.2005.06.007

[32] Morgado E, Gordon MK, Minana-Solis M d C, Meza E, Levine S, Escobar C, et al. Hormonal and metabolic rhythms associated with the daily scheduled nursing in rabbit pups. American Journal of Physiology. Regulatory and Integrative Comparative Physiology. 2008;**295**:R690-R695. DOI: 10.1152/ajpregu.00162.2008

[33] Morgado E, Meza E, Gordon MK, Pau FKY, Juárez C, Caba M. Persistence of hormonal and metabolic rhythms during fasting in 7-to-9-day-old rabbits entrained by nursing during night. Hormones and Behavior. 2010;**58**:465-472. DOI: 10.1016/j.yhbeh.2010.05.003

[34] Jilge B, Kuhnt B, Landerer W, Rest S. Circadian temperature rhythms in rabbit pups and their does. Laboratory Animal. 2001;**35**:364-373. DOI: 10.1258/0023677011911831

[35] Hudson R. Do newborn rabbits learn the odor stimuli releasing nipple-search behavior? Developmental Psychobiology. 1985;**18**:575-585. DOI: 10.002/dev.420180612

[36] Hudson R, Distel H. On the nature and action of the rabbit nipple-search pheromone: A review. In: Appfelbach R, Müller-Schwarze D, Reuter K, Weiler E, editors. Chemical Signals in Vertebrates VII. Oxford: Elsevier; 1995. pp. 223-232

[37] Fuchs AR, Cubile L, Dawood MY, Jørgensen FS. Release of oxytocin and prolactin by suckling in rabbits throughout lactation. Endocrinology. 1984;**114**:462-469. DOI: 10.1210/endo-114-2-462

[38] Bautista A, Mendoza-Degante M, Coureaud G, Martínez-Gómez M, Hudson R. Scramble competition in newborn domestic rabbits for an unusually restricted milk supply. Animal Behaviour. 2005;**70**:1011-1021. DOI: 10.1016/j.anbehav.2005.01.015

[39] Jilge B. Ontogeny of the rabbit's circadian rhythms without an external zeitgeber. Physiology & Behavior. 1995;**58**:131-140. DOI: 10.1016/0031-9384 (95)00006-5

[40] Meza E, Juárez C, Morgado E, Zavaleta Y, Caba M. Brief daily suckling shifts locomotor behavior and induces PER1 protein in paraventricular and supraoptic nuclei, but not in the suprachiasmatic nucleus of rabbit does. European Journal of Neuroscience. 2008;**28**:1396-1403. DOI: 10.1111/j.1460-9568.2008.06408.x

[41] Caba M, González-Mariscal G, Meza E. Circadian rhythms and clock genes in reproduction: Insights from behavior and the female rabbit's brain. Frontiers in Endocrinology. 2018;**9**:106. DOI: 10.3389/fendo.2018.00106

[42] González-Mariscal G, Jiménez A, Chirino R, Beyer C. Motherhood and nursing stimulate c-Fos expression in the rabbit forebrain. Behavioral Neuroscience. 2009;**123**:731-739. DOI: 10.1037/a0016487

[43] Moore-Ede MC, Sulzman FM, Fuller CA. The Clocks That Time Us: Physiology of the Circadian System. Cambridge (USA): Harvard University Press; 1982. p. 448. DOI: 10.1126/science.217.4563.925

[44] Davis FC. Ontogeny of circadian rhythms. In: Aschoff J, editor. Handbook of Behavioral Neurobiology. Vol. 4. New York: Plenum Press; 1981. pp. 257-274. DOI: 10:å1007/978-1-4615-6552-9

[45] Reppert SM, editor. Development of Circadian Rhythmicity and Photoperiodism in Mammals. Boston: Perinatology Press; 1989. p. 262

[46] Weinert D. Ontogenetic development of the mammalian circadian system. Chronobiology International. 2005;**22**:179-205. DOI: 10.1081/CBI-200053473

[47] Hudson R. Potential of the newborn rabbit for circadian rhythms research. Biological Rhythm Research. 1998;**29**:546-555. DOI: 10.1071/brhm.29.5.546.4828

[48] Allingham K, von Saldern C, Brennan PA, Distel H, Hudson R. Endogenous expression of c-Fos in hypothalamic nuclei of neonatal rabbits coincides with their circadian pattern of suckling-associated arousal. Brain Research. 1998;**783**:210-218. DOI: 10.1016/S0006-8993(97)01379-6

[49] Caldelas I, Tejadilla D, González B, Montúfar R, Hudson R. Diurnal pattern of clock gene expression in the hypothalamus of the newborn rabbit. Neuroscience. 2007;**144**:395-401. DOI: 10.1016/j.neuroscience.2006.09.020

[50] Caldelas I, González B, Montúfar-Chaveznava R, Hudson R. Endogenous clock gene expression in the suprachiasmatic nuclei of previsual newborn rabbits is entrained by nursing. Developmental Neurobiology. 2009;**69**:47-59. DOI: 10.1002/dneu.20687

[51] Montúfar-Chaveznava R, Hernández-Campos O, Hudson R, Caldelas I. Differential maturation of the molecular clockwork in the olfactory bulb and suprachiasmatic nucleus of the rabbit. Neuroscience. 2012;**207**:198-207. DOI: 10.1016/j.neuroscience.2012.01.025

[52] Hudson R, Müller A, Kennedy GA. Parturition in the rabbit is compromised by daytime nursing: The role of oxytocin. Biology of Reproduction. 1995;**53**:519-524. DOI: 10.1095/biolreprod53.3.519

[53] Ninomiya-Alarcón IG, Hudson R, Reyes-Guerrero G, Barrera-Mera B, Guevara-Guzman R. Effect of photoperiod on the mechanical response of the pregnant rabbit uterus to oxytocin. American Journal of Physiology. Regulatory and Integrative Comparative. 2004;**287**:R174-R180. DOI: 10.1152/ajpregu.00710.2003

[54] Apel S, Hudson R, Coleman GJ, Rödel HG, Kennedy GA. Regulation of the rabbit's once-daily pattern of nursing: A circadian or hourglass-dependent process? Chronobiology International. 2020;**37**:1151-1162. DOI: 10.1080/07420528.2020.1805459

[55] Wood DH. The demography of a rabbit population in an arid region of New South Wales, Australia. Journal of Animal Ecology. 1980;**49**:55-79. DOI: 10.2307/4277

[56] Palomares F. The negative impact of heavy rains on the abundance of a Mediterranean population of European rabbits. Mammalian Biology. 2003;**68**:224-234. DOI: 10.1078/1616-5047-00088

[57] Rödel HG, Starkloff A, Seltmann MW, Prager G, von Holst D. Causes and predictors of nest mortality in a European rabbit population. Mammalian Biology. 2009;**74**:198-209. DOI: 10.10167j.mambio.2008.04.003

[58] Findlay ALR, Roth LL. Long-term dissociation of nursing behavior and the condition of the mammary gland in the rabbit. Journal of Comparative and Physiological Psychology. 1970;**72**:341-344. DOI: 10.1037/h0029469

[59] Findlay ALR, Tallal PA. Effect of reduced suckling stimulation on the duration of nursing in the rabbit. Journal of Comparative and Physiological Psychology. 1971;**76**:236-241. DOI: 10.1037/h0031390

[60] Venge O. The influence of nursing behaviour and milk production on the early growth in rabbits. Animal Behaviour. 1963;**11**:500-506. DOI: 10.1016/0003-3472(63)90269-0

[61] González-Mariscal G, Lemus AC, Vega- González A, Aguilar-Roblero R. Litter size determines circadian periodicity of nursing in rabbits. Chronobiology International. 2013;**30**:711-718. DOI: 10.3109/07420528.2013.784769

[62] Rensing L, Meyer-Grahle U, Ruoff P. Biological timing and the clock metaphor: Oscillatory and hourglass mechanisms. Chronobiology International. 2001;**18**:329-369. DOI: 10.1081/CBI-100103961

Chapter 5

Biological Determinants of Sleep Disorders

Valery V. Gafarov, Elena A. Gromova, Vladimir N. Maksimov, Igor V. Gagulin and Almira V. Gafarova

Abstract

The purpose of the study is to research the effect of polymorphism of genes such as *CLOCK, ARNTL, PER2, NPAS2, DRD4, DAT, TNF-α,* and *NPSR1* on sleep disorders in an open population of 25–64-year-old men. We conducted screening studies of representative samples of men aged 25–64 years. The general examination was carried out according to the standard methods included in the WHO MONICA-Psychosocial Program (MOPSY). Carriers of the C/T genotype of the *CLOCK* gene more often than others reported having "satisfactory" or "poor" sleep. Carriers of the C/T genotype of the *ARNTL* gene were more likely to experience anxiety dreams, and they woke up exhausted. Carriers of the A/A genotype of the *PER2* gene were more likely to wake up two or more times per night, a total of four to seven times per week. In the population, C/T and T/T genotypes of the *NPAS2* gene were significantly more common in individuals with 7-hour sleep. Genotype 4/6 of the *DRD4* gene and genotype 9/9 of the *DAT* gene were significantly associated with sleep disturbances. Carriers of the heterozygous A/G genotype of the *TNF-α-308* gene, compared with carriers of all other genotypes, more often rated sleep as "satisfactory" (30%) than "good."

Keywords: population, men, sleep disorders, *CLOCK* gene, *ARNTL* gene, *PER2* gene, *NPAS2* gene, *DRD4* gene, *DAT* gene, *TNF-α* gene, *NPSR1* gene

1. Introduction

Sleep is a complex set of brain processes that support human physiological needs [1]. Sleep is part of the sleep-wake cycle. This cycle, consisting of approximately 8 hours of sleep at night and 16 hours of daytime wakefulness in humans, is controlled by a combination of two internal factors, that is, sleep homeostasis and circadian rhythms [2]. Unlike wakefulness, sleep is a period of inactivity and restoration of mental and physical functions. Sleep is thought to provide time for inputting information gained during waking into memory and for reestablishing communication between different parts of the brain. Sleep is also the time when other body systems replenish their energy and repair their tissues [3], and it is the key to wellness and optimal health [4–6]. People who get enough quality sleep have more energy, better cognitive function, memory, alertness, attention, and performance during the day, as well as a healthier immune system [7]. Quality healthy sleep is one of the basic needs of people and is important for their health [8].

Circadian rhythms are a system that synchronizes all processes in living organisms that provide temporary adaptation, including sleep and wakefulness. The study of circadian rhythms and biological clocks progressed slowly until the methods that anticipated the beginning of the genomic millennium came to the aid of scientists. At the end of the last century, scientists found out that there is a biological "clockwork mechanism" in the mammalian brain that coordinates the work of the entire organism. To be more precise, these clocks are located in the suprachiasmatic nucleus (SCN) of the hypothalamus. Today we know that each SCN neuron is a miniature clock counting the circadian rhythm, and all these thousands of clocks work in unison, forcing the rest of the body's systems to obey. The SCN receives information about the illumination from special receptors located on the retina of the eye and sends corresponding signals to other organs using hormones and nerve impulses. Some SCN cells, as well as cells of many other organs, have individual molecular clocks. The "gears" in these clocks are transcription factors, the activity of which changes over the day. The synthesis of several different proteins depends on the activity of these key transcription factors, which gives rise to the circadian rhythms of the vital activity of individual cells and entire organs. A bright light turned on early at night can shift the circadian rhythm, activating *PER* gene transcription, which usually occurs in the morning. However, it should be understood that a nerve impulse represents the final crescendo of the long processes unfolding in a neuron. To understand the nature of these processes, one has to descend from the cellular to the gene level [9, 10].

In the 1960s and early 1970s, Seymour Benzer at the California Institute of Technology [11] and Ronald Konopka, one of his students, studied the genetics of Drosophila behavior and the latter discovered the first circadian rhythm gene localized in the X chromosome [12]. The gene was named *period*, or *per* (the protein encoded by this gene was respectively named PER). Scientists have found three mutant alleles for *per*, in addition to the normal "wild type." With one of them, the daily cycle of the fly was shortened to 19 hours, with the other, it was lengthened to 29 hours, while the carriers of the third "did not observe hours" at all, that is, their periods of rest and activity were of random duration [12]. The 2017 Nobel Prize in physiology or medicine was awarded to Jeffrey C. Hall, Michael Rosbash, and Michael W. Young for cloning and sequencing the *period* gene in 1984 [13, 14]. Michael Rosbash and his colleagues also noticed that the concentration of messenger RNA (mRNA) of the *per* gene increases and decreases within 24 hours. In mutants, these fluctuations accelerated or slowed down [13]. In the 90s, *new details of the mechanism were discovered—the timeless* genes, or *tim* genes, *doubletime* (Michael W. Young) [15], as well as *Clock*, *cycle*, and *Cryptochrome* genes (Michael Rosbash group) [16].

The relationship between these genes and their products is seen in **Figure 1** [17].

The clock mechanism is based on two proteins—CLOCK (CLK) and BMAL1 (-ARNT-like protein 1 in the brain and muscle; also known as ARNTL or MOP3). By dimerizing, CLOCK and BMAL1 activate the transcription of the Period *(PER)* and Cryptochrome *(CRY)* genes. In nocturnal rodents, as well as in some diurnal animals, the transcription of the *PER1* and *PER2* genes in the SCN peaks in the morning or afternoon, while for *CRY1* and *CRY2* genes, the peak is observed closer to the evening. An increase in the concentration of PER and CRY proteins in the cell triggers a feedback mechanism that blocks further synthesis of these proteins. According to recent studies, the main inhibitor of the CLOCK-BMAL1 complex is CRY, but it works only when combined with PER. During the night, cellular enzymes gradually degrade

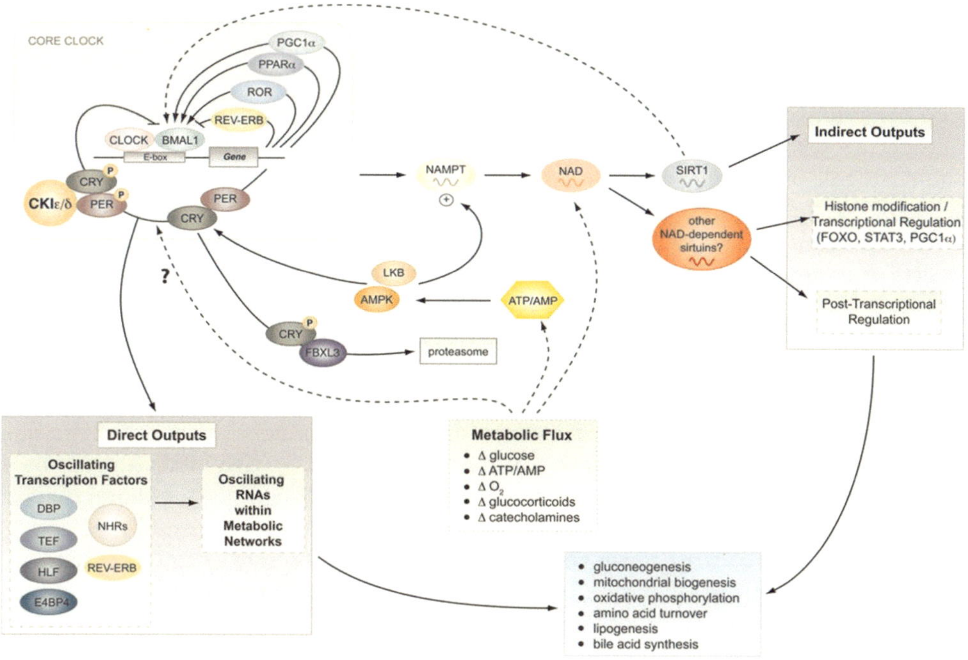

Figure 1.
Direct and indirect outputs of the core clock mechanism [17].

PER and CRY, and when their concentration reaches a critically low point, transcription is reactivated. The duration of the cycle depends on the degradation rate of PER and CRY [18, 19].

An important question is what molecular mechanisms provide a link between the light signal and the genes of the biological clock? Until recently, it was believed that phototransduction—the conversion of a light signal into an electrical signal transmitted through neurons—can only take place in the retina of the eye and only through the retinal, an active component of rhodopsin. In 2011, Fogle K.J., Parson K., et al. found that the CRY protein has the same ability, and it uses a mechanism independent of TIM and PER [20]. If the neuron in which CRY is expressed is illuminated by blue light, a complex cascade of reactions involving potassium membrane channels is triggered and an action potential is generated, that is, the neuron produces an electric signal under the direct influence of light. Control experiments have shown that this reaction has nothing to do with opsin, the visual pigment in Drosophila. However, if in the course of an experiment the other neurons not previously possessing photosensitivity are forced to synthesize CRY, they also start generating signals in response to flashes of light [20].

Mention should also be made of the *NPAS2* gene, which is located on chromosome 2 in 2q11.2, one of the most important circadian genes. NPAS2 forms heterodimers with BMAL1 and then activates the circadian genes *PER* and *CRY*, which are essential for maintaining biological rhythms in many organisms. Animal studies have shown that loss of normal function of the *NPAS2* gene can cause defects in several aspects of the circadian system, such as sleep patterns and behavior [21].

According to the current understanding of the neurophysiology of sleep, monoamines, one of which is dopamine, also play an important role [22]. Damage to central dopaminergic synaptic transmission plays an important role in the onset of severe neuropsychiatric disorders [23]. People with these conditions have serious sleep disturbances such as excessive daytime sleepiness [24], rapid eye movements disrupting sleep behavior [25], a decrease in the period of paradoxical sleep, and disrupted sleep architecture [26]. In general, all these observations suggest that dopamine plays an important role in the regulation of the sleep-wake cycle [27]. The evidence that increased extracellular dopamine is one of the key mechanisms of wakefulness activation also associates reduced dopamine metabolism with sleep disturbances [28].

TNF-α (tumor necrosis factor-) is a pro-inflammatory cytokine that contributes to the formation of atherosclerotic plaque. Although early reviews showed a contradictory association with CHD (coronary heart disease) [29], the replacement of G (guanidine) by A (adenine) at position 308 in the promoter region is now associated with increased TNF-α production [30], as well as with increased inflammatory response after cardiac surgery [31], insulin resistance [32], CHD, in individuals with type II diabetes [33] and increased C-reactive protein in individuals with vital exhaustion, defined as a condition with excessive fatigue, difficulty falling asleep, general malaise, apathy, irritability, and lack of energy [34]. In addition, it was suggested that variability in location 308 may affect the development of depressive symptoms, for example, allele A is more common in depressed patients than in controls [35]. There is an association between obstructive sleep apnea and the presence of the A allele [36].

The neuropeptide S receptor (NPSR1) is a metabotropic G protein receptor with seven transmembrane helices [37]. The receptor was first described in 2002, and in 2004 neuropeptide S (NPS) was identified as a ligand [38, 39]. NPS refers to neuropeptides, a diverse group of neurons expressing signaling molecules involved in various brain functions. Studies in rats have shown that administration of NPS greatly induces wakefulness and reduces the occurrence of all stages of sleep [38, 40]. NPS appears to be expressed only in certain brain regions [41–46]. The highest concentration of NPS precursor mRNA was found in brainstem neurons adjacent to the locus coeruleus, in the parabrachial nuclei of the variolar bridge, and the sensory nucleus of the trigeminal nerve [39, 40]. The locus coeruleus and parabrachial nuclei belonging to the ascending activating reticular system, as well as the sensory nucleus of the trigeminal nerve, are known for their contribution to modulating the sleep/wake cycle [42, 43]. Compared to NPS, the expression pattern of NPSR1, a precursor of mRNA, is more widely distributed in the brain. It covers important centers of the sleep/wake system in the hypothalamus and thalamus and is also present in the cortex and amygdala. In particular, it can be found in hypothalamic regions such as the peripheral region (red band) and the tuberomammillary nucleus, which are known for their expression of orexin and histamine, respectively [44–46]. In addition, NPSR1 mRNA was found in key regions responsible for sleep induction. At the molecular level, the receptor activates protein kinases and increases intracellular cAMP and Ca^{2+} levels [38]. Thus, NPS is believed to modulate the neurotransmission of the expressing neurons NPSR1. Although the NPS system appears to play a critical role in modulating sleep, most of the conclusions have been drawn from rodent studies, and data on its effect on sleep in humans are limited. A single rs324981 nucleotide polymorphism (lying at triplet position 107 of the *NPSR1* gene on chromosome 7p14.3) provides an opportunity for a noninvasive study of the effect of NPS/NPSR1 in the human body. The SNP T allele leads to the exchange of amino acids in the active center of the receptor binding site (Asn $\geq$ Ile) [47]. This, in turn, leads to an approximately tenfold increase in the sensitivity of neuropeptide S [47]. The T allele has already been identified as a risk factor for the development of asthma and panic disorder [48–50]. Gottlieb et al. [51] conducted a whole-genome sequencing that examined sleep parameters—falling asleep time and sleep duration. They found a connection between rs324981 and the time a person goes to bed. This study showed a delay in the moment of falling asleep among T-allele carriers.

Thus, our study aimed to investigate the effect of polymorphism of circadian rhythm genes (i.e., *CLOCK*, *ARNTL*, *PER2*, and *NPAS2*), dopamine receptor genes (i.e., *DRD4*, *DAT*), pro-inflammatory cytokines (*TNF-α* gene), and the *NPSR1* gene on sleep disorders in an open population of 25–64-year-old men.

2. Materials and methods

We conducted screening studies of representative samples of the population aged 25–64 years in one of the districts of Novosibirsk city (the budget theme No. AAAA17–117112850280-2):

at screening II in 1988–1989, 725 men were examined, average age—43.4 ± 0.4 years, the response rate was 71.3%;

at screening III in 1994–1995, 647 men were examined, average age—44.3 ± 0.4 years, the response rate was 82.1%;

at screening IV in 2003–2005, 576 men were examined, average age—54.23 ± 0.2 years, the response rate was 61%;

at screening V in 2013–2016, 427 men were examined, the average age—34 ± 0.4 years, the response rate was 71%;

at screening VI in 2016–2018, 275 men were examined, average age—49 ± 0.4 years, the response rate was 72%.

The general examination in 1988–1989, 1994–1995, 2003–2005, 2013–2016, and 2016–2018 was conducted according to standard methods included in the WHO MONICA-Psychosocial Program (MOPSY) [52].

A standard Jenkins questionnaire was used to study sleep disorders. Statistical analysis was performed using the SPSS software package version 20 [53].

Genotyping of the studied polymorphisms of circadian rhythm genes (*CLOCK, ARNTL, PER2, NPAS2* genes) (screening V), as well as those related to the dopaminergic system (*DRD4, DAT* genes) (screening III) and inflammatory pro-cytokines (*TNF-α*) (screening III) and *NPSR1* gene (screening IV), was performed in the Laboratory of Molecular Genetic Studies of the Research Institute of Therapy and Preventive Medicine—Branch of ICG SB RAS, Novosibirsk [Vladimir N. Maksimov, Doctor of Sciences (Medicine), is the head of the laboratory].

The distribution of traits and their quantitative characteristics was analyzed. Simple relations between variables were analyzed (contingency tables). The hypothesis of independence of factors A and B or homogeneity of factor B in relation to the levels of factor A was tested using the method of contingency table construction. The reliability of the independence of the factors was assessed by the criterion χ^2 [54]. Reliability was accepted at a significance level of $p \leq 0.05$.

3. Results

In the male population under study, the rate of sleep disturbances ("poor" and "very poor" sleep) in the group aged 25–34 years in 1988–1989 was 5.4%; in 1994–1995, it went down to 3.6%; and in 2013–2016, it reached 4.3%. In the group aged 35–44 years, the rate of sleep disorders in 1988–1989 was 9.5%; in 1994–1995, it was 9.3%; in 2013–2016, it decreased to 4.2%; and later in 2016–2018, it grew dramatically to 11%. In the group aged 45–54 years, the rate of sleep disorders was 11% in 1988–1989, 9.8% in 1994–1995, in 2003–2005, it rose to 12.5%, and fall to 4.9% in 2016–2018. Among men in the older age group of 55–64 years, the rate of sleep disturbance was the highest in 1988–1989—2020.8%; in 1994–1995, it plummeted to 12.1%; in 2003–2005, it showed a slight decrease to 11.8%; and finally, in 2015–2018, it rose to 19.7% (**Table 1**).

3.1 Association of polymorphism of the rs2412646 CLOCK gene with sleep disorders

Table 2 shows the frequency distribution of rs2412646 genotypes of the *CLOCK* gene among men aged 25–44 years in Novosibirsk. In the studied population of 25–44-year-old men, the most common was the homozygous C/C genotype of the *CLOCK* gene—50.3%, the heterozygous C/T genotype was found in 42.5% and the T/T genotype in only 7.2%.

The respondents were asked how well they sleep. Among carriers of the C/T genotype, the response "satisfactory" (36.8%) and "poorly" (5.3%) sounded more often than among carriers of all other genotypes ($\chi^2 = 9.44$ df = 4 $p < 0.05$). Carriage of the T/T and C/T genotype of rs2412646 *CLOCK* gene was most frequently combined with carriage of the A/A genotype of rs934945 *PER2* gene among men aged 25–44 years in Novosibirsk (30.8% and 46.2%, respectively). The carriers of the C/C genotype of rs2412646 *CLOCK* gene most often had A/G and G/G genotypes of rs934945 *PER2* gene (68.4% and 68.9%, respectively) ($\chi^2 = 27.18$ df = 4 $p = 0.001$).

3.2 Association of the polymorphism of the rs2278749 ARNTL gene with sleep disorders

Table 3 shows the frequency distribution of rs2278749 genotypes of the *ARNTL* gene among 25–44-year-old men in Novosibirsk.

The most common rs2278749 genotype of the *ARNTL* gene was the homozygous C/C genotype—74.9%, the second most common was the heterozygous C/T genotype—22.3%, while only 2.8% of individuals in the population had the homozygous T/T genotype.

The question was asked: "How often in the last month have you had disturbing dreams while sleeping?" Only C/T genotype carriers experienced such disturbances for 22 days or more, that is, 7.5%; 27.5% of C/T genotype and 20% of T/T genotype carriers reported having disturbing dreams 4–7 days per month, while C/C genotype carriers more often than others responded that they had no disturbing dreams at all, that is, 42.5% ($\chi^2 = 16.35$ df = 8 $p < 0.05$).

The question, "During the past month, how often did you wake up after a normal sleep, feeling tired or exhausted?", showed that 40% of T/T genotype carriers experienced such problems one to three times a month, while C/T genotype carriers (7.5%) experienced this problem more often than others, from 15 days to 3 weeks ($\chi^2 = 18.71$ df = 8 $p < 0.01$). Men carrying the heterozygous C/T genotype (57.3%) were more likely than others to wake up feeling exhausted and tired ($\chi^2 = 19.52$ df = 4 $p = 0.001$).

Question/Attitude: Do you sleep well? Rate the quality of your sleep.		25–34		35–44		45–54		55–64		25–64*	
Screening		N	%	N	%	N	%	N	%	N	%
Screening II 1988–1989	1. Very good	35	17.2	16	8	10	5.8	8	5.4	69	9.5
	2. Good	101	49.5	91	45.7	59	34.1	42	28.2	293	40.5
	3. Satisfactory	57	27.9	73	36.7	85	49.1	68	45.6	283	39
	4. Poor	10	4.9	17	8.5	15	8.7	30	20.1	72	9.9
	5. Very poor	1	0.5	2	1	4	2.3	1	0.7	8	1.1
	Total	204	100	199	100	173	100	149	100	725	100
	χ^2 = 68,611 df = 12 p < 0.001										
Screening III 1994–1995	1. Very good	9	5.4	11	6.4	8	5.6	8	4.8	36	5.6
	2. Good	97	58.1	72	41.9	55	38.5	75	45.5	299	46.2
	3. Satisfactory	55	32.9	73	42.4	66	46.2	62	37.6	256	39.6
	4. Poor	5	3	14	8.1	13	9.1	17	10.3	49	7.6
	5. Very poor	1	0.6	2	1.2	1	0.7	3	1.8	7	1
	Total	167	100	172	100	143	100	165	100	647	100
	χ^2 = 20,148 df = 12 p < 0.001										
Screening IV 2003–2005	1. Very good					10	3.3	10	3.7	20	3.5
	2. Good					104	34.2	73	26.8	177	30.7
	3. Satisfactory					152	50	157	57.7	309	53.6
	4. Poor					37	12.2	29	10.7	66	11.5
	5. Very poor					1	0.3	3	1.1	4	0.7
	Total					304	100	272	100	576	100
	χ^2 = 5720 df = 4 p > 0.05										
Screening V 2013–2016	1. Very good	25	15.2	28	10.7					53	12.4
	2. Good	78	47.6	126	48.3					205	48
	3. Satisfactory	54	32.9	96	36.8					151	35.4
	4. Poor	6	3.7	10	3.8					16	3.7
	5. Very poor	1	0.6	1	0.4					2	0.5
	Total	164	100	261	100					427	100
	χ^2 = 2200 df = 4 p > 0.05										
Screening VI 2016–2018	1. Very good			6	8.3	3	3.7	0	0	9	3.3
	2. Good			32	44.4	35	43.2	35	28.7	102	37.1
	3. Satisfactory			26	36.1	39	48.1	63	51.6	128	46.5
	4. Poor			8	11.1	4	4.9	24	19.7	36	13.1
	5. Very poor			0	0	0	0	0	0	0	0
	Total			72	100	81	100	122	100	275	100
	χ^2 = 24.636 df = 6 p < 0.001										

*Note *—for screening IV, 45–64 years old, for screening V, 25–44 years old, for screening VI, 35–64 years old.*

Table 1.
Prevalence of sleep disorders in an open population of males 25–64 years of age for the period 1988–2018.

rs2412646 genotypes of the *CLOCK* gene	**N**	**%**
T / T	13	7.2
C/T	76	42.5
C/C	90	50.3
Total	179	100
The rs2412646 alleles of the *CLOCK* gene	N	%
T	102	28.5
C	256	71.5
Total	358	100
Hardy-Weinberg equilibrium	χ^2 = 0.0409, p = 0.7344 q = 0.2656	

Table 2.
rs2412646 polymorphism of the CLOCK *gene in a population of men aged 25–44 years (screening V).*

Genotypes rs2278749 of the *ARNTL* gene	**N**	**%**
T / T	5	2.8
C/T	40	22.3
C/C	134	74.9
Total	179	100
Allele rs2278749 of the *ARNTL* gene	N	%
T	50	14
C	308	86
Total	358	100
Hardy-Weinberg equilibrium	χ^2 = 10.134 p = 0.8641 q = 0.1359	

Table 3.
Polymorphism rs2278749 of the ARNTL *gene in an open population of men aged 25–44 years (screening V).*

3.3 Association of the polymorphism of the rs934945 PER2 gene with sleep disorders

The prevalence of polymorphic variants of candidate gene of rs934945 *PER2* gene was as follows—homozygous genotype A/A—4.47%, heterozygous genotype A/G—30.17%, and homozygous genotype G/G—65.36%. **Table 4** shows the frequency distribution of the rs934945 *PER2* gene among 25–44-year-old men in Novosibirsk.

Among individuals with the homozygous A/A genotype of the *PER2* gene, there was a growing tendency for anxiety dreams during sleep a total of 4–7 days per month (12.5%), compared with A/G genotype carriers (7.4%) and G/G genotype carriers (12%) (χ^2 = 13.83 df = 8 p = 0.08).

The carriers of the G/G genotype of the *PER2* gene were significantly less likely to wake up at night (51.9%) compared to men carrying the A/G (38.5%) and A/A (37.5%) genotypes. In contrast, men carrying the AA genotype were more likely (25%) to wake up (two or more times per night), a total of four to seven times per week, compared to the carriers of the A/G genotype (20.4%) and the G/G genotype (13.2%) (χ^2 = 25.76 df = 8 p = 0.004).

Genotypes of rs934945 *PER2* gene	N	%
A/A	8	4.47
A/G	54	30.17
G/G	117	65.36
Total	179	100
Alleles rs934945 of the *PER2* gene	N	%
A	62	17.3
G	296	82.7
Total	358	100
Hardy-Weinberg equilibrium	χ^2 = 0.4852 p = 0.8128 q = 0.1872	

Table 4.
Polymorphism of rs934945 PER2 *gene in an open population of men aged 25–44 years (screening V).*

In the population of men aged 25–44 years, individuals carrying the A/A genotype of the *PER2* gene tended to have a shorter sleep duration of 5 hours or less (62.5%) compared to carriers genotypes A/G (57.4%) and G/G (41.9%) (χ^2 = 9.21 df = 10 p = 0.51).

3.4 Association of the polymorphism of the rs934945 PER2 gene with sleep disorders

The prevalence of polymorphic variants of the candidate gene NPAS2 rs4851377 in the population was as follows—homozygous C/C genotype in 13.4%, heterozygous C/T genotype in 53.6%, and homozygous genotype T/T in 32.9%. The C allele of the NPAS2 candidate gene was found in 40.3% of men, and the T allele was found in 59.7% of men (**Table 5**).

In the population, the C/C genotype of the NPAS2 rs4851377 gene was significantly more common in those who slept at least 8 hours per night (33.3%) and 9 hours per night (33.3%), and the C/T and T/T genotypes were in those with 7 hours of sleep (50% and 53.3%, respectively) (χ^2 = 18.425 df = 10 $p < 0.05$).

It was found that 6-hour sleep 4.5 times (95% CI 0.735–27.536) was significantly more often observed among the carriers of the T allele than the C allele who had 9-hour sleep (χ^2 = 6.142 df = 1 $p < 0.05$); also 7-hour sleep versus 9-hour sleep was four times more often (95% CI 0.66–24.245) in carriers of the T allele than in carriers of the C allele (χ^2 = 5.488 df = 1 $p < 0.05$).

3.5 Association of VNTR polymorphisms of the DRD4 gene and VNTR polymorphism of the DAT gene with sleep disorders (screening III)

In an open population of men aged 25–64 years, the frequency of homozygous genotype 4/4 of the dopamine receptor subtype 4 (*DRD4*) gene was 57.9%; genotype 2/2 was less frequent—6.1%, genotype 2/4—12.5%, and genotype 3/4—5.6%; even less frequent were—genotype 4/6—4.2%, genotype 2/6, genotypes 4/7 and 6/6 were present in equal proportions—2.1%. The frequency distribution of alleles showed that allele 4 prevails—found in 70.7%, allele 2 was found in 14%, allele 6—in 6%. The remaining alleles account for 0.8%–5.4% (**Table 6**).

Genotypes of rs4851377 *NPAS2* gene	**N**	**%**
C/C	24	13.4
C/T	96	53.6
T/T	59	32.9
Total	179	100
Alleles of rs4851377 *NPAS2* gene	N	%
C	144	40.3
T	214	59.7
Total	358	100
Hardy-Weinberg equilibrium	χ^2 = 0.2221 p = 0.6453 q = 0.3547	

Table 5.
Polymorphism of rs4851377 NPAS2 *gene in an open population of men aged 25–44 years (screening V).*

The sleep self-assessment distribution of VNTR carriers of *DRD4* gene polymorphism is presented in **Table 6**. Carriers of genotype 2/4 of the *DRD4* gene were more likely to respond that their sleep was "good" (16.2%) than "satisfactory" (7.9%), both compared with carriers of all other genotypes (χ^2 = 5.626 υ = 1 $p < 0.01$) and compared with carriers of genotype 4/4 (χ^2 = 4.507 υ = 1 $p < 0.05$). Self-assessment of sleep as "poor/good" revealed a significant difference between carriers of genotypes 2/4 and 4/6; in the former, self-assessment of sleep as "good" predominated (16.2%), and the latter more often assessed their sleep as "poor" (28.6%) (χ^2 = 5.714 df = 1, $p < 0.05$). For the "good/very good" sleep self-assessment, the "very good" (15.4%) rather than "good" (4.3%) responses were more common among genotype 3/4 carriers (χ^2 = 5.199 υ = 1 $p < 0.05$). Among carriers of genotype 4/4, sleep is more often assessed as "satisfactory" (61.8%) than "poor" (38.1%), as in comparison with carriers of all other genotypes (χ^2 = 7.687 υ = 1, $p < 0.01$), and in comparison with carriers of genotype 4/6 OR = 12.7 (95% CI 4.1–38.7); (χ^2 = 26.941 υ = 1 $p < 0.0001$). A reliable result was obtained when comparing carriers of genotype 4/4, in which the score "good sleep" prevails (60%), with carriers of genotype 2/6, in which the score "poor sleep" prevails (7.1%), OR = 10, 4 (95% CI 1.6–67.1); (χ^2 = 8.772 df = 1 $p < 0.01$). Self-assessment of sleep as "good" (60%) in carriers of genotype 4/4 is more common than assessment as "very good" (40%) among carriers of all other genotypes (χ^2 = 6.664 υ = 1 $p < 0.01$) and in comparison with carriers of genotype 2/4 (15.4%) (χ^2 = 5.223 υ = 1 $p < 0.05$). Among carriers of genotype 4/6, there is an increase in responses of "poor sleep" (28.6%) rather than "satisfactory" (3.6%) in comparison with carriers of all other genotypes OR = 10.6 (95% CI 3.6–30.4); (χ^2 = 26.217 υ = 1 $p < 0.0001$); with carriers of genotype 2/4 OR = 5.2 (95% CI 1.2–21.5); (χ^2 = 5.461 υ = 1 $p < 0.05$).

Carriers of allele 2 of the *DRD4* gene more often assessed their sleep as "good" (15.4%) than "poor" (9.5%), in comparison with carriers of all other alleles (χ^2 = 5.739 υ = 1 $p < 0, 01$) as well as with carriers of allele 3 (χ^2 = 4.790 υ = 1 $p < 0.05$). Carriers of allele 4 were more "satisfied" (71.5%) with their sleep than carriers of all other alleles (χ^2 = 4.101 υ = 1 $p < 0.05$). Carriers of allele 5 rated their sleep as "very poor" (7.1%) much more often than "satisfactory" (0.9%), in comparison with carriers of all other alleles OR = 8.3 (95% CI 0.8–86.1); (χ^2 = 4.541 υ = 1 $p < 0.05$).

Genotype	Population		Sleep self-assessment									
			Very good		Good		Satisfactory		Poor		Very poor	
	n	%	n	%	n	%	n	%	n	%	n	%
2/2	26	6.1	3	11.5	11	5.9	11	6.7	0	0	1	14.3
2/3	1	0.2	0	0	0	0	1	0.6	0	0	0	0
2/4	53	12.5	4	15.4	30	16.2	13	7.9	5	11.9	1	14.3
2/5	2	0.5	0	0	2	1.1	0	0	0	0	0	0
2/6	10	2.4	0	0	2	1.1	5	3	3	7.1*	0	0
2/7	1	0.2	0	0	1	0.5	0	0	0	0	0	0
3/3	8	1.9	0	0	4	2.2	4	2.4	0	0	0	0
3/4	24	5.6	4	15.4*	8	4.3	8	4.8	4	9.5	0	0
3/6	3	0.7	0	0	1	0.5	1	0.6	1	2.4	0	0
3/7	2	0.5	0	0	0	0	2	1.2	0	0	0	0
4/4	246	57.9	13	40	111	60	102	61.8***	16	38.1	4	57.1
4/5	4	0.9	0	0	1	0.5	1	0.6	1	2.4	1	14.3
4/6	18	4.2	0	0	0	0	6	3.6	12	28.6**	0	0
4/7	9	2.1	0	0	6	3.2	3	1.8	0	0	0	0
4/8	1	0.2	0	0	0	0	1	0.6	0	0	0	0
5/5	3	0.7	0	0	2	1.1	1	0.6	0	0	0	0
5/6	2	0.5	2	7.7	0	0	0	0	0	0	0	0
6/6	9	2.1	0	0	5	2.7	4	2.4	0	0	0	0
7/7	3	0.7	0	0	1	0.5	2	1.2	0	0	0	0
χ^2 = 161.067 df = 72 p = 0.0001												
Allele	**Population**		**Very good**		**Good**		**Satisfactory**		**Poor**		**Very poor**	
	n	%	n	%	n	%	n	%	n	%	n	%
2	26	6.1	10	19.2	57	15.4	41	12.4	8	9.5	3	2.4
3	9	2.1	4	7.7	17	4.6	20	6.1	5	6.0	0	0
4	323	76.0	34	65.4	267	72.2	236	71.5	54	64.3	10	71.4
5	9	2.1	2	3.8	7	1.9	3	0.9	1	1.2	1	7.1
6	42	9.9	2	3.8	13	3.5	20	6.1	16	19.0	0	0
7	15	3.5	0	0	9	2.4	9	2.7	0	0	0	0
8	1	0.2	0	0	0	0	1	0.3	0	0	0	0
χ^2 = 46,555 df = 24 p = 0.004												
-p<0,05; **-p<0,01; *-p<0,001*												

Table 6.
Frequencies of genotypes and alleles of VNTR polymorphism of the DRD4 *gene in the population and their association with sleep disorders (screening III).*

3.6 Carriers of allele 6 of the *DRD4* gene more often rated their sleep as "poor" (19%)

a. rather than "satisfactory" (6.1%) in carriers of all other alleles OR = 3.6 (95% CI 1.7–7.4); (χ^2 = 14.224 υ = 1 $p < 0.0001$) and in comparison with carriers of allele 2 (χ^2 = 8.097 υ = 1 $p < 0.004$), allele 3 (χ^2 = 3.905 υ = 1 $p < 0.05$), allele 4 (χ^2 = 12.665 υ = 1 p < 0, 00001);

b. rather than "good" (3.5%) compared to carriers of all other alleles OR = 6.4 (95% CI 2.9–14); (χ^2 = 27.626 υ = 1 $p < 0.0001$); carriers of allele 2 (χ^2 = 19.379 υ = 1 $p < 0.0001$), allele 3 (χ^2 = 5.437 υ = 1 $p < 0.05$), allele 4 (χ^2 = 5.048 υ = 1 $p < 0.05$), and allele 5 (χ^2 = 5.147 υ = 1 p = 0.05);

c. and also rather "very good" (3.8%) in comparison with carriers of all other alleles OR = 5.8 (95% CI 1.2–26.7); (χ^2 = 6.463 υ = 1 $p < 0.01$) (**Table 6**).

In the frequency distribution of VNTR genotypes of the *DAT* gene polymorphism in a population of men aged 25–64 years with different self-assessed sleep, no significant differences were found. Positive sleep assessments were more common—41.9% of carriers of genotype 9/10 had "good" sleep, among carriers of genotype 10/10, "very good" sleep was observed in 61.5%. Among the carriers of the 9/9 genotype, 8.3% considered their sleep to be "poor" (χ^2 = 25.486 υ = 32 $p > 0.05$). The distribution of the remaining carriers of various *DAT* genotypes and self-assessment of sleep did not exceed 3.8% (**Table** 7).

There were no significant differences in the frequency distribution of the *DAT* gene alleles in a population of men aged 25–64 years, with different self-assessment of sleep. There was a tendency for an increase in positive responses 78.8% of carriers of allele 10 reported having "very good" sleep, while 26.4% carriers of allele 9 assessed their sleep as "poor." The distribution of the other alleles of the *DAT* gene did not exceed 3.8% (χ^2 = 16.777 υ = 20 $p > 0.05$) (**Table** 7).

3.7 Association of G308A polymorphism of TNF- α gene with sleep disorders

The frequency of genotypes G308A polymorphism of the tumor necrosis factor *TNF-α* gene in the male population of Novosibirsk city is presented in **Table 8**. The G/G genotype of the *TNF-α* gene was found in 79.1% of individuals, the A/G genotype was found in 19% of cases, and finally, the A/A genotype was only found in 1.9% of men. In the population, 88.6% of men had allele G and only 11.4% had allele A (**Table 8**).

There was a tendency that the G/G genotype of the *TNF-α* gene was more common not only in individuals with very good (88.2%) and good (83.3%) sleep but also in the group with sleep disorders: poor (84.2%) and very poor (100%) sleep. The genotype A/G of the *TNF-α* gene was most often found among men who consider their sleep "satisfactory" (30%). The A/A genotype of the *TNF-α* gene did not exceed 5.3% among all categories of sleep disorders (χ^2 = 12,012 df = 8 p = 0,151).

There was also a tendency toward a more frequent occurrence of the G allele of the *TNF-α* gene among all categories of men with different assessments of their sleep, and the A allele of the *TNF-α* gene in the group of men who rate their sleep as satisfactory (17.5%) (χ^2 = 9.451 df = 4 p = 0.051).

Comparative analysis in groups with various self-assessment of sleep showed that among carriers of genotype G/G gene *TNF-α*, in comparison with carriers of all other genotypes, assessment of sleep as "good" (98.3%) rather than satisfactory (67.5%) χ^2 = 36,943 df = 1 p = 0.0001; OR = 27,685 (95% CI 6339–120,906), and "good" (98.3%), followed by "poor" (84.2%) χ^2 = 9151 df = 1 p = 0.02; OR = 10,781

Genotype	Sleep self-assessment											
	Population		Very good		Good		Satisfactory		Poor		Very poor	
	n	%	n	%	n	%	n	%	n	%	n	%
8/8	4	1	1	3.8	1	0.6	2	1.2	0	0	0	0
9/9	15	3.7	0	0	2	1.1	10	6.2	3	8.3	0	0
6/10	3	0.7	0	0	1	0.6	2	1.2	0	0	0	0
8/10	1	0.2	0	0	1	0.6	0	0	0	0	0	0
9/10	149	36.6	9	34.6	75	41.9	51	31.7	13	36.1	1	20
10/10	223	54.8	16	61.5	90	50.3	94	58.4	19	52.8	4	80
10/11	4	1.0	0	0	3	1.7	1	0.6	0	0	0	0
10/12	1	0.2	0	0	1	0.6	0	0	0	0	0	0
11/11	7	1.7	0	0	5	2.8	1	0.6	1	2.8	0	0
χ^2 = 25.486 df = 32 p = 0.786												
Allele	**Sleep self-assessment**											
	Population		Very good		Good		Satisfactory		Poor		Very poor	
	n	%	n	%	n	%	n	%	n	%	n	%
6	3	0.4	0	0	1	0.3	2	0.6	0	0	0	0
8	9	1.1	2	3.8	3	0.8	4	1.2	0	0	0	0
9	179	22	9	17.3	79	22.1	71	22	19	26.4	1	10
10	604	74.2	41	78.8	261	72.9	242	75.2	51	70.8	9	90
11	18	2.2	0	0	13	3.6	3	0.9	2	2.8	0	0
12	1	0.1	0	0	1	0.3	0	0	0	0	0	0
χ^2 = 16.777 df = 20 p = 0.667												

Table 7.
Frequencies of VNTR genotypes and alleles of the DAT *gene polymorphism in the population and their association with sleep disorders (screening III).*

(95% CI 1672–69,537) is much more common. On the contrary, among carriers of the heterozygous A/G genotype of the *TNF-α* gene, in comparison with carriers of all other genotypes, sleep was more often satisfactory (30%) than good (15.2%) χ^2 = 6756 df = 1 p = 0.009. A comparative analysis between the carriers of different genotypes of *TNF-α* helped to reveal that the carriers of the genotype G/G reported good sleep (84.6%) rather than satisfactory (69.2%), more often, whereas in carriers of genotype A/G, on the contrary, satisfactory sleep (30.8%), rather than good sleep (15.4%) χ^2 = 7013 df = 1 p = 0,008; OR = 2434 (95%CI 1247–4751) prevails. Comparative analysis of carriers of the G and A alleles of the *TNF-α* gene among men with different self-assessment of sleep did not show significant differences.

3.8 Association of polymorphism of the rs324981 NPSR1 gene with sleep disorders (screening IV)

In an open population of men aged 45–64 years, the frequency of homozygous C/C genotype of the *neuropeptide S* gene (*NPSR1* rs324981 gene) was 19.4%, with a lower

***TNF-α-308G/A* genotype**	**n**	**%**
G/G	204	79.1
A/G	49	19
A/A	5	1.9
Total	258	100
Allele of *TNF-α-308G/A* gene	n	%
G	457	88.6
A	59	11.4
Total	516	100
Hardy-Weinberg equilibrium	χ^2 = 1.0964, p = 0.8878 q = 0.1122	

Table 8.
Tumor necrosis factor TNF-gene α *G308A polymorphism (screening III).*

***NPSR1* rs324981 genotype**	**n**	**%**
C/C	7	19.4
C/T	19	52.8
T/T	10	27.8
Total	36	100
The allele of the *NPSR1* gene rs324981	n	%
C	33	45.8
T	39	54.2
Total	72	100
Hardy-Weinberg equilibrium	χ^2 = 0.1859, p = 0.6163 q = 0.3832	

Table 9.
NPSR1 *gene rs324981 polymorphism in 45–64-year-old male population (screening IV).*

frequency of T/T genotype—27.8%, the most common in men was the heterozygous C/T genotype—52.8%. The frequency distribution of alleles showed that among men the T-allele prevails—54.2% and C-allele was found in 45.8% of individuals (**Table 9**).

An associative analysis of the allele frequencies of the neuropeptide S polymorphism of the *NPSR1* rs324981 gene in men with different self-assessment of sleep revealed that T-allele carriers more frequently rated their sleep as "satisfactory"—in 69% of cases, while C-allele carriers did it only in 57.1% of cases. In addition, carriers of the C-allele are more often satisfied with their sleep, rating it as "good"—28.6% and "very good" than carriers of the T-allele who have "good" sleep only in 20.7% of cases (χ^2 = 15,713 df = 8, $p < 0.05$).

4. Discussion

Sleep plays an important role in promoting health. Studies conducted over the past decade have confirmed that sleep disturbance has a powerful effect on the risk of infectious diseases, the occurrence and progression of some major diseases, including cardiovascular diseases and cancer, as well as the incidence of depression [55, 56]. In Russia, about 45% of adults are dissatisfied with their sleep, and almost 20% need serious treatment for sleep disorders [57]. In our population, the level of sleep disorders turned out to be high and had the following trend—it demonstrated a decrease from 1988 to 1989 to 1994–1995 and an increase in 2003–2018. The increase in sleep disturbance in 2003–2018 was observed mainly in the older age groups (45–64 years old). Problems with sleep in the population can only get worse every year. The rapid emergence of "24/7" communities, working 24 hours a day, 7 days a week, participation in round-the-clock events, increased nighttime use of television, the internet, and mobile phones, means that an adequate uninterrupted nighttime sleep is becoming rare. The proportion of workers with circadian rhythm disturbances who are needed to service "24/7" communities is likely to increase [58], which determined the feasibility of studying biological determinants of sleep disorders.

The *CLOCK* gene encoding the positive transcription factor CLOCK is among the major circadian rhythm genes. CLOCK protein with binding partner BMAL1, *BMAL1* gene product, forms a transactivation dimer influencing the promoter of controlled genes [59]. In our study, we examined the associative relationship between sleep disorders and various polymorphic variants of the rs2412646 genotype of the *CLOCK* gene among men aged 25–44 years. The response of having "satisfactory" sleep (36.8%) and "poor" sleep (5.3%) was found to be more frequent among C/T genotype carriers than among carriers of all other genotypes (χ^2 = 9.44 df = 4 $p < 0.05$). The frequency distribution of rs2412646 genotypes of the *CLOCK* gene depending on the rs934945 genotype of the *PER2* gene is also of interest. We found that carriers of the rs2412646 genotype of the *CLOCK* C/T gene were most often carrying the A/A rs934945 genotype of the *PER2* gene in the male population of 25–44 years old. Our results are supported by the findings obtained by other researchers regarding the association of the *CLOCK* gene with insomnia [60, 61] and preference for a particular sleep-wake cycle [62–64].

The *ARNTL* gene is a key element of the positive feedback loop of the molecular circadian oscillator [65]. According to the activity of some genes, Jun Z. Li, et al. 2013 [66] were able to determine that in individuals suffering from affective disorders circadian rhythms are interrupted, and "night" genes were expressed during the day. It has been suggested that desynchronization may occur due to a disruption in the connection between individual circadian genes. Considering the association of genotypes of the rs2278749 *ARNTL* gene with affective disorders, we have identified several components that have shown the strongest association with polymorphic variants of the gene under study. It turned out that carriers of the genotype C/T had problems with sleep, especially since they had much more anxiety dreams during the month. Men, carriers of the T allele, both homozygotes (T/T genotype) and heterozygotes (C/T genotype) were more likely to wake up tired or exhausted. Thus, our results confirm the data of other researchers concerning the identification of individual polymorphic variants of the *ARNTL* gene, leading to possible desynchronization and disruption of the circadian rhythm and, accordingly, leading to affective disorders [66].

Examining the association of *PER2* gene genotypes with sleep characteristics, we found that among carriers of the A/A genotype there is a tendency to have more anxiety dreams in comparison with carriers of other genotypes. Moreover, A/A genotype carriers were more likely to wake up during the night, and the tightest sleep was observed in men who were A/G and G/G genotype carriers. Sleep deprivation (5 hours or less) also occurred more frequently in individuals whose genotype contained the homozygous A allele. Our results overlap in part with those obtained by Ojeda D.A., et al. [67], who studied the association of the *PER2* gene (rs934945) with circadian rhythms in healthy individuals, students at Columbia University. The *PER2* gene (rs934945) showed a statistically significant association with two subscales of the morning sleepiness scale, that is, "activity planning" and "morning alertness." The association of rs934945 with "morning restlessness" was first shown.

The most common in the population was the heterozygous genotype of the candidate gene *NPAS2 C/T*—53.3%, followed by the homozygous genotype T/T, with both variants of the candidate gene more common among men who had enough sleep for only 7 hours a day. The C/C genotype of the *NPAS2* candidate gene was significantly more common in those who slept for at least 8 hours (33.3%) or 9 hours (33.3%) per day. The major T allele of the candidate gene *NPAS2* was 4.5 times more common in men who sleep 6 hours a day and four times more common in men with 7 hours of sleep. Thus, the obtained data indicate that rs4851377 of the candidate NPAS2 gene determines whether men are night owls or early birds [21].

According to the literature devoted to genetic research, it has been established that some mental and emotional characteristics of a person are associated with polymorphism of the 3rd exon of the gene of the neurotransmitter system of the dopamine receptor, subtype 4 (*DRD4*) [68] and the dopamine transporter gene (*DAT*) [69].

When considering the occurrence of identified polymorphic variants of *DRD4* candidate genes in people with sleep disorders, it was found that carriers of the genotype 4/4 were more likely to believe that they had either good or satisfactory sleep. Carriers of genotypes 2/4 and 3/4 were more likely to rate their sleep positively, while men with genotypes 2/6 and 4/6, on the contrary, were dissatisfied with their sleep and rated it as "poor" more often.

When comparing the "short" and "long" alleles of the *DRD4* gene, we observed approximately the same pattern: carriers of the "long" allele 6 more often evaluated their sleep as "poor," carriers of allele 2 believed that their sleep was "very good," and carriers of the most common in the population allele 4 mostly reported having "good" sleep.

According to the current understanding of the biosynthesis of dopamine, it is known that just one sleepless night is enough for its level in the brain to increase [70]. The findings of the study conducted by Nora Volkow et al. suggest that dopamine in the human brain is involved in the so-called adaptation process, which leads to sleep disturbance. The researchers also found that the amount of dopamine in the brain is associated with feelings of fatigue and physical ability to perform cognitive tasks. However, the study also found that increased levels of dopamine in the brain cannot compensate for cognitive disorders caused by lack of sleep. On the other hand, according to the literature, people with the "long" form of the *DRD4* gene (six or more repeat units) have a lowered affinity of dopamine receptors and a reduced number of receptors. These individuals are less sensitive to dopamine. This means that they need more stimulation to get the same reaction than people with a "short" gene [71]. It can be assumed that "stimulating wakefulness," for example, a sleepless night, is one of the "ways" to naturally raise the level of dopamine, to receive a "reward" by the brain, for which we later have to pay with insomnia.

When analyzing the frequency distribution according to the conjugation tables of genotypes and alleles of the VNTP polymorphism of the *DAT* gene in a population of 25–64-year-old men, with different sleep self-assessment, no significant differences were found. There was only a tendency toward an increased number in the positive assessments of sleep-in carriers of genotype 9/10 and genotype 10/10 of the *DAT* gene. Negative sleep assessments were slightly more frequent in carriers of genotype 9/9 of the *DAT* gene. There was a tendency for an increase in the number of those who reported their sleep to be "very good" among the carriers of allele 10 and "poor" among the carriers of allele 9. Dopamine uptake is carried out through active transmembrane transfer using the dopamine protein transporter, it has been experimentally established that disabling the *DAT* gene in mice leads to a reduction in the paradoxical sleep phase (REM sleep) and promotes early awakening [28]. In individuals containing a short variant of the *DAT* gene in the genome, the reuptake of dopamine is altered [72], and there is reason to believe that an increase in free dopamine contributes to an increase in the period of wakefulness, but, as mentioned above, does not contribute to either physical or mental rest [70]. Probably, for this reason, men carrying the genotype 9/9 of the *DAT* gene are *more* likely to evaluate their sleep negatively.

In the Novosibirsk population, the most frequent genotype-G308A of the tumor necrosis factor *TNF-α* gene polymorphism was the G/G genotype—it was observed in almost 80% of men. Moreover, the same genotype was predominant in all groups differing in sleep quality. Comparative analysis showed that individuals with homozygous genotype G /G are much more likely to give positive assessments of their sleep, unlike carriers of all other genotypes combined, who evaluate sleep as satisfactory or even poor. In addition, heterozygous carriers with the A/G genotype are less likely to positively assess sleep than carriers of the homozygous G/G genotype. Although there are no direct analogous studies in the world literature, nevertheless, studies devoted to the study of obstructive sleep apnea can provide indirect evidence. According to various authors, insomnia occurs in 42%–54.9% of patients with sleep breathing disorders [73]. A meta-analysis published in 2012 showed an association between 308G/A and obstructive sleep apnea—the presence of the 308 A allele increases the risk of developing obstructive sleep apnea by 65%, compared with individuals with a homozygous G/G genotype ($OR = 1.65$, 95% $CI = 1.02–2.68$, $p = 0.04$) [36].

In the study population, we found an association between some polymorphic variants of the *NPSR1* gene and sleep self-assessment among men. The presence of the T allele in the genotype was found to contribute to poorer sleep quality among men. Our results are confirmed by the results obtained by Spada J et al. [74], who proved that the study participants with the homozygous T/T genotype had significantly shorter sleep/rest times than individuals carrying the C allele in the genotype. These findings are confirmed by the studies by G.W.A. Gottlieb et al. [51], who found associations between rs324981 and sleep. Our conclusion about the relationship between sleep quality and rs324981 is also consistent with studies on rats since the direct introduction of NPS into the rat brain strongly affects sleep architectonics. Already within the first hour after injecting NPS into the brain, wakefulness time was lengthened, while the rapid eye movement phase, as well as the slow sleep phase, were shortened [39]. Similar results were obtained by Zhao et al. [40], who reported a decrease in sleep phases. These studies indicate that NPS can inhibit different sleep phases that were previously thought to be independently regulated [75]. However, the causal relationship can be much more complicated, as it is still unknown how rs324981 polymorphism acts in ontogeny. There is a hypothesis that various compensatory mechanisms may be induced simultaneously with the loss of allele function, or there may be interaction with other unknown genetic or environmental factors, which may explain the lack of association with sleep disturbances in heterozygotes [76, 77].

5. Conclusions

1. It was found that 25–64-year-old men scored high in sleep disturbance. We observed the following trend—firstly, a decline from 1988 to 1989 to 1994–1995 (11% and 8.6%, respectively) and later growth in 2003–2018 (13.1%). The increase in sleep disturbance in 2003–2018 was observed mainly in the older age groups (45–64 years old).

2. The most common genotype in the population was C/C rs2412646 of the *CLOCK* gene (50.3%), C/T was found in 42.5%, and the genotype T/T in 7.2%. Carriers of the C/T genotype of the *CLOCK* gene more often than others reported having "satisfactory" or "poor" sleep. Carriage of the rs2412646 genotype of the *CLOCK* T/T and C/T gene was most often combined with carriage of the A/A rs934945 genotype of the *PER2* gene. Carriers of the C/C rs2412646 genotype of the *CLOCK* genotype most often had the A/G and G/G rs934945 genotypes of the *PER2 gene.*

3. The most common genotype rs2278749 of the *ARNTL* gene was the homozygous C/C genotype (74.9%), the C/T genotype was found in 22.3% of individuals, and 2.8% were the carriers of the T/T genotype. Carriers of the C/T genotype were more likely to experience anxiety dreams, and they woke up tired or exhausted; on the contrary, carriers of the C/C genotype were much less likely to experience anxiety dreams.

4. The prevalence of polymorphic variants of the candidate gene *rs934945* of the *PER2* gene was as follows—genotype A/A—4.47%, genotype A/G—30.17%, and genotype G/G—65.36%. Among individuals with the A/A genotype of the *PER2* gene, there was a tendency of seeing a larger number of anxiety dreams for a total of 4–7 days per month (12.5%), compared with carriers of the A/G genotypes (7.4%) and carriers of the G/G genotype (12%). G/G carriers of the *PER2* genotype were significantly less likely to wake up at night (51.9%), and men, carriers of the genotype A/A, on the contrary, woke up more often (25%) twice or more times per night, in general, from four to seven times a week. Individuals carrying the genotype A/A of the *PER2* gene tended to have a shorter sleep duration of 5 hours or less (62.5%), compared to carriers of the genotypes A/G (57.4%) and G/G.

5. The prevalence of polymorphic variants of the candidate gene *NPAS2* rs4851377 was as follows—C/C genotype was found in 13.3%, C/T genotype—in 53.3%, and T/T genotype—in 33.3%. In the population, the C/C genotype of the *NPAS2* rs4851377 gene was significantly more common in those who slept at least 8 hours a day (33.3%) and 9 hours (33.3%), and the C/T and T/T genotypes were found in people with 7 hours of sleep (50% and 53.3%, respectively). It was found that 6-hour sleep was 4.5 times significantly more often observed among the carriers of the T allele than among the carriers of the C allele, who had 9-hour sleep. Moreover, 7-hour sleep versus 9-hour sleep was four times more often found in carriers of the T allele than in carriers of the C allele.

6. In the male population, homozygous genotype 4/4 of the dopamine receptor of subtype 4 *DRD4* gene is the most represented (57.9%). With the frequency distribution of VNTR genotypes of *DAT* gene polymorphism, it was found that

the homozygous genotype 10/10 is more common (54.8%), whereas the heterozygous genotype 9/10 was less common (36.6%). Genotype 9/9 was observed in 3.7%. The incidence of other genotypes was 1.7% or less. Genotype 4/6 of the *DRD4* gene and genotype 9/9 of the *DAT* gene were significantly associated with sleep disorders.

7. The G/G genotype of the *TNF-α-308* gene was found in 79.1% of individuals, the A/G genotype in 19% of cases, and the A/A genotype was found in 1.9% of men. In the population, 88.6% of men had the G allele and 11.4% had the A allele. Carriers of the G/G genotype of the *TNF-α-308* gene compared to carriers of all other genotypes assessed their sleep as "good" (98.3%) much more often than as "satisfactory" (67.5%) or "poor" (84.2%). Carriers of the heterozygous A/G genotype of the *TNF-α-308* gene, compared with carriers of all other genotypes, more often rated sleep as "satisfactory" (30%) than "good" (15.2%).

8. The frequency of the homozygous C/C genotype of the neuropeptide S gene (*NPSR1* rs324981 gene) was 19.4%, T/T was less frequent (27.8%), and the most common was the heterozygous C/T genotype (52.8%). A tendency of growing dissatisfaction with the quality of their sleep among men with T/T genotype carriers has been revealed—70%. Male T-allele carriers were more likely to report sleep disturbances than C-allele carriers.

Acknowledgements

The research was carried out within the framework of budget theme No. AAAA-A17-117112850280-2.

Conflict of interest

No conflict of interest.

Author details

Valery V. Gafarov[1,2], Elena A. Gromova[1,2]*, Vladimir N. Maksimov[3], Igor V. Gagulin[1,2] and Almira V. Gafarova[1,2]

1 Laboratory of Psychological and Sociological Issues of Internal Diseases, Institute of Internal and Preventive Medicine - Branch of Institute of Cytology and Genetics SB RAS, Novosibirsk, Russia

2 Collaborative Laboratory of Cardiovascular Diseases Epidemiology, Novosibirsk, Russia

3 Laboratory of Molecular Genetic Investigation of Internal Diseases, Institute of Internal and Preventive Medicine - Branch of Institute of Cytology and Genetics SB RAS, Novosibirsk, Russia

*Address all correspondence to: elena.a.gromova@gmail.com

References

[1] Kirszenblat L, van Swinderen B. The yin and yang of sleep and attention. Trends in Neurosciences. 2015;**38**: 776-786

[2] Phillips B, Gelula R. Sleep-wake cycle: Its physiology and impact on health. National Sleep Foundation. 2006:1-9

[3] Carol M, Glenn PM. Pathophysiology: Concepts of Altered Health States. 8th ed. New York, NY, USA: Wolters Kluwer Health Lippincott Williams & Wilkins; 2009

[4] Sleep Counsel. The Great British Bedtime Report. Technical Report. 2013. Available from: http://www.sleepcouncil.org.uk/wp-content/uploads/2013/02/The-Great-British-Bedtime-Report.pdf.

[5] Holfeld B, Ruthig JC. A longitudinal examination of sleep quality and physical activity in older adults. Journal of Applied Gerontology. 2014;**33**(7): 791-807

[6] Watson NF, Badr MS, Belenky G, et al. Recommended amount of sleep for a healthy adult: A joint consensus statement of the American Academy of Sleep Medicine and Sleep Research Society. Journal of Clinical Sleep Medicine. 2015;**11**(6):591-592

[7] Bils P. Quality sleep: The center of a healthy life, evidence of the essential role of sleep — and what happens when we don't get enough of it. 2017. pp. 1-31. Available from: https://sleepnumber.new-media-release.com/360smartbed/downloads/SleepNumber_SleepWhitePaper_FINAL.pdf

[8] Åkerstedt T. Psychosocial stress and impaired sleep. Scandinavian Journal of Work, Environment & Health. 2006; **32**(6):493-501

[9] Green CB, Takahashi JS, Bass J. The meter of metabolism. Cell. 2008;**134**: 728-742

[10] Suter DM, Schibler U. Feeding the Clock. Science. 2009;**326**:378-379

[11] Anderson D, Brenner S. Seymour Benzer (1921–2007). Nature. 2008;**451**: 139. DOI: 10.1038/451139a

[12] Konopka RJ, Benzer S. Clock mutants of Drosophila melanogaster. Proceedings of the National Academy of Sciences of the United States of America. 1971;**68**(9):2112-2116. DOI: 10.1073/pnas.68.9.2112

[13] William PR, David AZ, Wheeler A, Pirrotta V, Hadfield C, Hall JC, et al. Molecular analysis of the *period* locus in Drosophila melanogaster and identification of a transcript involved in biological rhythms. Cell. 1984;**38**(3): 701-710

[14] Young MW, Jackson FR, Shin H-S, Bargiello TA. A biological clock in *Drosophila*. Cold Spring Harbor Symposia on Quantitative Biology. 1985; **50**:865-875. DOI: 10.1101/SQB.1985.050.01.104

[15] Vosshall LB, Young MW. Circadian rhythms in drosophila can be driven by period expression in a restricted group of central brain cells. Neuron. 1995;**15**(2): 345-360

[16] Emery P, VenusSo W, Kaneko M, Hall JC, Rosbash M. CRY, a *Drosophila* clock and light-regulated cryptochrome, is a major contributor to circadian rhythm resetting and photosensitivity. Cell. 1998;**95**(5):669-679

[17] Bass J, Takahashi JS. Circadian integration of metabolism and

energetics. Science. 2010;**330**(6009): 1349-1354. DOI: 10.1126/science.1195027

[18] Lamia KA, Sachdeva UM, DiTacchio L, Williams EC, Alvarez JG, Egan DF, et al. AMPK regulates the circadian clock by cryptochrome phosphorylation and degradation. Science. 2009;**326**:437-440

[19] Busino L, Bassermann F, Maiolica A, Lee C, Nolan PM, Godinho SI, et al. SCFFbxl3 controls the oscillation of the circadian clock by directing the degrada tion of cryptochrome proteins. Science. 2007;**316**:900-904

[20] Foglekelly KJ, Parsonnicole G, Todd DA, Holmes C. Cryptochrome is a blue-light sensor that regulates neuronal firing rate. Science. 2011;**331**(6023): 1409-1413. DOI: 10.1126/science.1199702

[21] Dudley CA, Erbel-Sieler C, Estill SJ, et al. Altered patterns of sleep and behavioral adaptability in NPAS2-deficient mice. Science. 2003;**301**:379-383

[22] Monti JM, Jantos H. The roles of dopamine and serotonin, and of their receptors, in regulating sleep and waking. Progress in Brain Research. 2008;**172**:625-646

[23] Greenwood TA, Schork NJ, Eskin E, et al. Identification of additional variants within the human dopamine transporter gene provides further evidence for an association with bipolar disorder in two independent samples. Molecular Psychiatry. 2006;**11**:125-133

[24] Abbott A. Neuroscience – While you were sleeping. Nature. 2005;**437**: 1220-1222

[25] Gagnon JF, Bedard MA, Fantini ML, et al. REM sleep behavior disorder and REM sleep without atonia in Parkinson's disease. Neurology. 2002;**59**:585-589

[26] O'Brien LM, Ivanenko A, Crabtree VM, et al. Sleep disturbances in children with attention deficit hyperactivity disorder. Pediatric Research. 2003;**54**: 237-243

[27] Dzirasa K, Ribeiro S, Costa R, et al. Dopaminergic control of sleep–wake states. The Journal of Neuroscience. 2006;**26**:10577-10589

[28] Wisor JP, Nishino S, Sora I, et al. Dopaminergic role in stimulant-induced wakefulness. The Journal of Neuroscience. 2001;**21**:1787-1794

[29] Allen RA, Lee EM, Roberts DH, et al. Polymorphisms in the TNF-alpha and TNF-receptor genes in patients with coronary artery disease. European Journal of Clinical Investigation. 2001; **31**:843-851

[30] Wilson AG, Symons JA, McDowell TL, et al. Effects of a polymorphism in the human tumor necrosis factor alpha promoter on transcriptional activation. Proceedings of the National Academy of Sciences of the United States of America. 1997;**94**:3195-3199

[31] Tomasdottir H, Hjartarson H, Ricksten A, et al. Tumor necrosis factor gene polymorphism is associated with enhanced systemic inflammatory response and increased cardiopulmonary morbidity after cardiac surgery. Anesthesia and Analgesia. 2003;**97**: 944-949

[32] Dalziel B, Gosby AK, Richman RM, et al. Association of the TNF-alpha −308 G/A promoter polymorphism with insulin resistance in obesity. Obesity Research. 2002;**10**:401-407

[33] Vendrell J, Fernandez-Real JM, Gutierrez C, et al. A polymorphism in the promoter of the tumor necrosis factor-alpha gene (−308) is associated

with coronary heart disease in type 2 diabetic patients. Atherosclerosis. 2003; **167**:257-264

[34] Jeanmonod P, von Kanel R, Maly FE, Fischer JE. Elevated plasma C-reactive protein in chronically distressed subjects who carry the A allele of the TNF-alpha −308 G/A polymorphism. Psychosomatic Medicine. 2004;**66**: 501-506

[35] Jun TY, Pae CU, Hoon H, et al. Possible association between -G308A tumour necrosis factor-alpha gene polymorphism and major depressive disorder in the Korean population. Psychiatric Genetics. 2003;**13**:179-181

[36] Huang J, Liao N, Huang QP, Xie ZF. Association between tumor necrosis factor-α-308G/A polymorphism and obstructive sleep apnea: A meta-analysis. Genetic Testing and Molecular Biomarkers. 2012;**16**(4):246-251

[37] Pitti T, Manoj N, Uversky VN. Molecular evolution of the neuropeptide S receptor. PLoS One. 2012;**7**(3):e34046

[38] Xu Y, Reinscheid RK, Huitron-Resendiz S, Clark SD, Wang Z, et al. Neuropeptide S. Neuron. 2004;**43**(4): 487-497

[39] Sato S, Shintani Y, Miyajima N, Yoshimura K. Novel G protein-coupled receptor protein and DNA thereof. World Patent Application. 2002;**WO 02/ 31145**:A1

[40] Zhao P, Shao YF, Zhang M, Fan K, Kong XP, et al. Neuropeptide S promotes wakefulness through activation of the posterior hypothalamic histaminergic and orexinergic neurons. Neuroscience. 2012;**207**:218-226

[41] Reinscheid R. Neuropeptide S: Anatomy, pharmacology, genetics and physiological functions. In: Civelli O, Zhou Q, editors. Orphan G Protein-Coupled Receptors and Novel Neuropeptides. Berlin Heidelberg: Springer; 2008. pp. 145-158

[42] Cairns BE, Fragoso MC, Soja PJ. Activity of rostral trigeminal sensory neurons in the cat during wakefulness and sleep. Journal of Neurophysiology. 1995;**73**(6):2486-2498

[43] Kohlmeier KA, Soja PJ, Kristensen MP. Disparate cholinergic currents in rat principal trigeminal sensory nucleus neurons mediated by M1 and M2 receptors: A possible mechanism for selective gating of afferent sensory neurotransmission. The European Journal of Neuroscience. 2006;**23**(12): 3245-3258

[44] Sutcliffe JG, De Lecea L. The hypocretins: Setting the arousal threshold. Nature Reviews Neuroscience. 2002;**3**(5):339-349

[45] Xu Y, Gall CM, Jackson VR, Civelli O, Reinscheid RK. Distribution of neuropeptide S receptor mRNA and neurochemical characteristics of neuropeptide S-expressing neurons in the rat brain. The Journal of Comparative Neurology. 2007;**500**(1): 84-102

[46] Jones BE. Arousal systems. Frontiers in Bioscience. 2003;**8**:438-451

[47] Reinscheid RK, Xu Y-L, Okamura N, Zeng J, Chung S, Pai R, et al. Pharmacological characterization of human and murine neuropeptide S receptor variants. The Journal of Pharmacology and Experimental Therapeutics. 2005;**315**(3):1338-1345

[48] Domschke K, Reif A, Weber H, Richter J, Hohoff C, et al. Neuropeptide S receptor gene — Converging evidence

for a role in panic disorder. Molecular Psychiatry. 2010;**16**(9):938-948

[49] Glotzbach-Schoon E, Andreatta M, Reif A, Ewald H, Troger C, et al. Contextual fear conditioning in virtual reality is affected by 5HTTLPR and NPSR1 polymorphisms: Effects on fear-potentiated startle. Frontiers in Behavioral Neuroscience. 2013;**7**:31

[50] Laas K, Reif A, Akkermann K, Kiive E, Domschke K, et al. Interaction of the neuropeptide S receptor gene Asn107Ile variant and environment: contribution to affective and anxiety disorders, and suicidal behavior. The International Journal of Neuropsychopharmacology. 2014;**17**(4):541-552

[51] Gottlieb DJ, O'Connor GT, Wilk JB. Genome-wide association of sleep and circadian phenotypes. BMC Medical Genetics. 2007;**8Suppl**:1S9

[52] World Health Organization. MONICA Psychosocial Optional Study Suggested Measurement Instruments. Copenhagen: WHO Regional Office for Europe; 1988

[53] Jenkins CD, Stanton BA, Niemeryk SJ, Rose RM. A scale for the estimation of sleep problems in clinical research. Journal of Clinical Epidemiology. 1988; **41**:313-321

[54] Bühl A, Zöfel P. SPSS Version 10. Einführung in die moderne Datenanalyse unter Windows. Saint Petersburg: DiaSoft; 2005. p. 608

[55] Besedovsky L, Lange T, Haack M. The sleep-immune crosstalk in health and disease. Physiological Reviews. 2019; **99**(3):1325-1380. DOI: 10.1152/physrev.00010.2018

[56] Robbins R, Affouf M, Seixas A, Beaugris L, Avirappattu G, Jean-Louis G. Four-year trends in sleep duration and quality: A longitudinal study using data from a commercially available sleep tracker. Journal of Medical Internet Re search. 2020;**22**(2):e14735. DOI: 10.2196/14735

[57] Wayne AM, Levin YaI. Insomnia. Clinical Medicine. 1998;**8**:53-55

[58] Ford ES, Cunningham TJ, Croft JB. Trends in self-reported sleep duration among US adults from 1985 to 2012. Sleep. 2015;**38**(5):829-832. DOI: 10.5665/sleep.4684

[59] Schantz V. Phenotypic effects of genetic variability in human clock genes on circadian and sleep parameters. Journal of Genetics. 2008;**87**(5):513-519

[60] Benedetti F, Dallaspezia S, Fulgosi MC, et al. Actimetric evidence that *CLOCK* 3111 T/C SNP influences sleep and activity patterns in patients affected by bipolar depression. American Journal of Medical Genetics Part B: Neuropsychiatric Genetics. 2007;**144B**: 631-635

[61] Serretti A, Benedetti F, Mandelli L, et al. Genetic dissection of psychopathological symptoms: Insomnia in mood disorders and *CLOCK* gene polymorphism. American Journal of Medical Genetics Part B: Neuropsychiatric Genetics. 2003;**121B**: 35-38

[62] Katzenberg D, Young T, Finn L, et al. A *CLOCK* polymorphism associated with human diurnal preference. Sleep. 1998;**21**:569-576

[63] Mishima K, Tozawa T, Satoh K, et al. The 3111T/C polymorphism of h*Clock* is associated with evening preference and delayed sleep timing in a Japanese population sample. American Journal of Medical Genetics Part B:

Neuropsychiatric Genetics. 2005;**133**: 101-104

[64] Friedman L, Zeitzer JM, Kushida C, et al. Scheduled bright light for treatment of insomnia in older adults. Journal of the American Geriatrics Society. 2009;**57**:441-452

[65] Hosoda H, Motohashi J, Kato H, Masushige S, Kida S. A *BMAL1* mutant with arginine 91 substituted with alanine acts as a dominant-negative inhibitor. Gene. 2004;**338**:235-241

[66] Li JZ, Bunney BG, Meng F, Hagenauer MH, Walsh DM, et al. Circadian patterns of gene expression in the human brain and disruption in major depressive disorder neuroscience. Proceedings of the National Academy of Sciences of the United States of America. 2013;**110**(24):9950-9955

[67] Ojeda DA, Perea CS, Niño CL, Gutiérrez RM, López-León S, Arboleda H, et al. A novel association of two non-synonymous polymorphisms in *PER2* and *PER3* genes with specific diurnal preference subscales. Neuroscience Letters. 2013;**11**(553):52-56

[68] Cloninger CR. A systematic method for clinical description and classification of personality variants: A proposal. Archives of General Psychiatry. 1987;**44**: 573-588

[69] Van Tol HHM, Bunzow JR, Guan HC, Sunahara RK, Seeman P, Niznik HB, et al. Cloning of the gene for a human dopamine *D4* receptor with high affinity for the antipsychotic clozapine. Nature. 1991;**350**:610-614

[70] Volkow ND, Wang GJ, Telang F, Fowler JS, Logan J, Wong C, et al. Sleep deprivation decreases binding of [^{11}C] raclopride to dopamine D_2/D_3 receptors in the human brain. Journal of Neuroscience. 2008;**28**(34):8454-8461. DOI: 10.1523/JNEUROSCI.1443-08.2008

[71] Ebstein RP et al. Additional evidence of an association between the dopamine *D4* dopamine receptor (*D4DR*) exon III repeat polymorphism and the human personality trait of novelty seeking. Molecular Psychiatry. 1997;**2**(6):472-477

[72] Gerra G, Garofano L, Pellegrini C, Bosari S, Zaimovic A, Moi G, et al. Allelic association of a dopamine transporter gene polymorphism with antisocial behaviour in heroin-dependent patients. Addiction Biology. 2005;**10**(3):275-281

[73] Beneto A et al. Comorbidity between sleep apnea and insomnia. Sleep Medicine Reviews. 2009;**13**:287-293

[74] Spada J, Sander C, Burkhardt R, Häntzsch M, Mergl R, Scholz M, Hegerl U, Hensch T. Genetic association of objective sleep phenotypes with a functional polymorphism in the *neuropeptide S* receptor gene. PLoS One. 2014;9(6):e98789. DOI: 10.1371/journal.pone.0098789

[75] Brown RE, Basheer R, McKenna JT, Strecker RE. McCarley RW control of sleep and wakefulness. Physiological Reviews. 2012;**92**(3):1087-1187

[76] Klauke B, Deckert J, Zwanzger P, Baumann C, Arolt V, et al. Neuropeptide S receptor gene (*NPSR*) and life events: G×E effects on anxiety sensitivity and its subdimensions. The World Journal of Biological Psychiatry. 2012;**15**(1):17-25

[77] Gibson G. Decanalization and the origin of complex disease. Nature Reviews. Genetics. 2009;**10**(2):134-140

Chapter 6

Chronotherapy Advances in the Management of Chronic Neurological and Cardiovascular Diseases: Complex Interactions of Circadian Rhythm Environmental Inputs, Nutrition and Drug Administration and Their Impact on Human Health

Alina Crenguța Nicolae, Ion-Bogdan Dumitrescu, Camelia Cristina Diaconu, Mirela Elena Ritivoiu, Carmen Adella Sirbu and Cristina Manuela Drăgoi

Abstract

New scientific evidence raises awareness concerning the human-specific interplay among primary environmental conditions, such as the light–dark cycle, activity–rest alternation, nutritional patterns, and their reflection on the physiological and pathological characteristics that are displayed uniquely by every individual. One of the critical aspects in the clinic is to understand the role of circadian rhythms as remarkable modulators of the biological effects of drugs and to aim for an optimal overlapping of the time of administration of medicines with the physiologic release of certain hormones, the time-dependent expression of genes, or the key-regulatory protein synthesis, which are all circadian-driven processes. The pharmacokinetics and pharmacodynamics profiles, as well as the possible drug interactions of neurotropic and cardiovascular agents, are intensely subjected to endogenous circadian rhythms, being essential to identify as much as possible the patients' multiple risk factors, from age and gender to lifestyle elements imprinted by dietary features, sleep patterns, psychological stress, all the way to various other associated pathological conditions and their own genetic and epigenetic background. This review chapter will highlight the involvement of biological rhythms in physiologic processes and their impact on various pathological mechanisms, and will focus on the nutritional impact on the circadian homeostasis of the organism and neurologic and cardiovascular chronotherapy.

Keywords: biological rhythms, chronopharmacology, chronotherapy, chrononutrition, cardiovascular drugs, neurologic diseases, sleep disorders

1. Introduction

Chronopharmacology, in its broadest sense, is the science that studies drug effects according to the time of their administration. As the human organism portrays a set of biological rhythms, its response to drug administration depends on its particular conditions at the moment when the drug enters the system. Indeed, the existence of rhythmic variations in the body circumstances reflects in the response to drugs: chronopharmacology studies these phenomena by assessing the variations in the activity, toxicity, and kinetics of medicines. Chronotherapy is the application of chronopharmacology outcomes, whose aim is to improve the benefit/risk ratio of the drug by optimizing the time of administration [1–3].

Many physiological processes in the array of metabolic balance, hormones synthesis, and release and nevertheless the sleep–wake behavior, are regulated by the circadian clock system, being closely related to everyday environmental inputs, such as the light–dark cycle, food consumption, and drug administration [4].

2. Chronobiology's importance from the perspective of human health

Chronopharmacology is a punctual aspect of chronobiology, reflecting the variations in the activity or toxicity of a therapeutic agent according to the time of administration, but it also studies the modifications of biological rhythms as the length of their cycle and time of their greatest and weakest intensity due to drugs. Its goal is to improve therapeutic efficacy and reduce unsolicited effects [1].

Relevant to metabolic activity, chronopharmacology allows to preserve or improve the health state, while associating circadian modifications in the digestive, hepatic, and endocrine systems with the exact time of the patient's meals. Studies show that sleep disorders or jet lag can be treated with drugs due to their beneficial impact on the adjustment of the circadian system. The connection between food and the circadian clock system has recently been encompassed in the term chrononutrition (**Figure 1**) [5, 6].

3. Human circadian rhythms, entrainment mechanisms, and major regulatory pathways

For humans, the most prevalent circadian rhythm is represented by the sleep–wake cycle, being at its turn in close relationship with the light–dark pattern, imprinted by environmental circumstances.

The central pivotal role in generating and maintaining basal circadian rhythms is played by the suprachiasmatic nucleus (SCN) located at the level of the anterior hypothalamus, which is the most important circadian pacemaker responsible for establishing the physiological cycles and nevertheless behavioral and endocrine circadian patterns displayed by human beings. The SCN function is highly important for daily rhythms that, when it is surgically removed, as has been experimentally performed in rodent studies, the animals lose their ability to temporally synchronize with the environment [3].

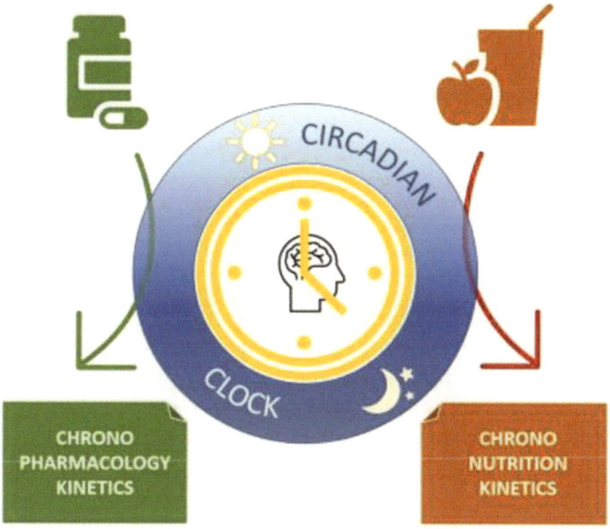

Figure 1.
Illustration of chronopharmacological and chrononutritional interactions with human organism and behavior.

The accumulated forfeits of different harmful events resident at the level of SCN essentially encountered in end-stage neurodegenerative diseases, translate into losing the sense of time: patients are going to sleep during the day or are remaining in a wakeful state in the dark period, they are feeling permanently hungry or going to the toilet at random intervals during the day or night [7, 8].

At this level, a link is established between the outer world and the interior one, considering the mechanisms of entrainment of the organism by sensing the light signals and transposing them into temporal information, and regulatory patterns disseminated to all downstream effector organs, tissues, and cells. The circadian regulatory pathway has remarkable profound echoes, implementing changes even in time-dependent genes, which modulate the expression of protein synthesis, resetting all other biological clocks present in the pineal gland, the pituitary gland, or the adrenal gland. Besides light, there are different important zeitgebers that coordinate the hepatic or gastrointestinal tract clocks, mainly by the time of meals or by drug administration. The flexibility of these internal clocks is mandatory, in terms of connecting the SCN signals to the hunger and satiety center located also in the hypothalamus, instructing the relevant organs to act at the solicited time. After a desynchronizing event, the downstream reverberations are perceived slowly, but firmly, and the system resets itself within a few days, for example, midnight nutrients ingestion resets the gut and liver clocks in less than a week, aligning the metabolic machinery to the new consumption pattern [9–12].

4. Phase shift modulation in the context of disruptive versus resynchronizing events

The human circadian structures are conceptualized in three distinctive components: a circadian oscillator with a rhythm of about 24 hours, pathways for the perception of light and other stimuli that synchronize the pacemaker with the environment, through zeitgebers, and effector systems and proper activities determined by SCN refinement. In humans, light is the dominant synchronizing agent for the internal clock, the photic information being conducted by direct and indirect pathways to the circadian system. In addition to the photic signal, the suppression of melatonin synthesis at the level of the pineal gland is a complementary mechanism implicated in conducting phase shifts in human circadian rhythms [12, 13].

The circadian shifts have an impact on activities including drug or nutrient absorption, distribution, metabolism, and excretion. Acknowledging these parameters when prescribing a drug and establishing the posology, dose and individual chemical characteristics, or the interactions with food intake can help improve human health and disease by increasing the potency of pharmacological and functional food effects. Secondly, just like light stimulation, drugs and food may be used to alter the phase of circadian clocks [14–16]. Internal clock disruptive events can occur simply by changing the timing of meals. Accordingly, the term chrononutrition also includes the following two elements: the involvement of food components and meal timing in the preservation of the health state and the role of food components in rapidly changing or reorganizing the endogenous clocks [10].

The importance of well-established nutritional routines, considering mainly the time of meals and the dietary habits, is highly acknowledged by all research studies in the field of nutrition. The reveal of the biochemical mechanism by which biological clocks are operating, namely the negative feedback regulation for the transcriptional process by means of binding Clock/Bmal1 to the E-box, was solid proof for pleading for stable eating habits. Circadian rhythms involve a clock regulated by negative transcriptional feedback. Clock/Bmal1, transcription factors, bind to E-box hexanucleotides to activate transcription of Per and Cry clock genes. The complex formed by Per and Cry inhibits its transcriptional activation by Clock/Bmal1. Subsequently, decreased activation of Per and Cry in turn causes transcriptional activation. This cycle takes approximately 24 hours. Although small gaps appear between cells, these gaps are adjusted by synchronizers [3, 17, 18].

This stability assures a certain equilibrium in the regulation of lipid metabolism, but more importantly, regulates the expression of the liver clock genes and the CYP7A1 isoform [4, 14, 16].

As such, chrononutrition will become a standard technique for maintaining our health via circadian rhythm system modulation. Knowing all the mechanistic details of food entrainment of the circadian clock will support the development of chrononutritional approaches for assuring nutritional optimal functionality.

5. The involvement of nutritional inputs and patterns in maintaining the circadian homeostasis of the organism

It is now understood that all cells have their own autonomous 24-hour clocks that work together as organ clocks, collectively forming a factor-integrated clock that synchronizes all organs. The rhythm of the digestive system is reversed when meal times are reversed, indicating that the digestive system clock synchronizer is sometimes stronger than the primary stimulus, namely light. It has been concluded by certain researchers that meals are among the strongest synchronizers for all organs and systems [19, 20].

The clocks of the organs cooperate to control the functions of the whole body, which can be defined as healthy. Experiments using mice with the clock gene removed show that its loss causes not only behavioral but also metabolic disorders. Experimental studies on circadian clock mutant mice exhibiting obesity or metabolic syndrome received much attention. In addition to arrhythmia, which was originally predicted, metabolic disorders in mice revealed that the circadian clock is strongly linked to peripheral metabolism [7, 21, 22]. On the other hand, there is a report of familial advanced sleep phase syndrome due to Per2 mutations in humans.

Gastric and intestinal digestion and absorption follow a circadian rhythm, being affected by clock genes rhythmically produced in the intestine and by the daily food intake (**Figure 2**). Extensive research has been conducted on the circadian expression of clock genes in the gastrointestinal tract. The obtained results revealed an advanced phase characteristic for the upper gut, being entrained faster than the lower gut, which is translated into a modification of the nutritional delivery rate at this level [17, 23].

The gastrointestinal system is subjected to a series of influence factors that are under tight circadian control, performed through genes exhibiting circadian patterns, which in their turn represent a signaling pathway. The most pronounced influence is driven by the nutritional schedule, being able to advance the idea of nutritional signaling. The colonic motility displays a clear rhythm, being active during the light period of the day and extremely silent during the night, the nitric oxide synthase activity and the clock genes regulating this cyclic process. The intestinal enzymes that perform the digestion of nutrients at this level have a circadian modulated activity timetable, which intimately follows the meal ingestion patterns, predicting the ideal time for synthesizing or activating these molecules. Consequent to this, the intestinal absorption of nutrients and xenobiotics registers a circadian fluctuation in close relationship with specific transporters' expression [16, 17, 24].

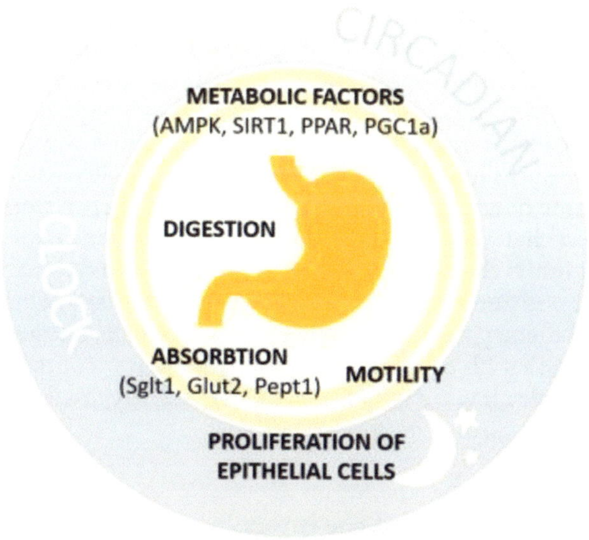

Figure 2.
Circadian rhythmicity of nutrition-related processes regulators. AMPK, 5′ adenosine monophosphate-activated protein kinase; SIRT1, sirtuin 1 or silent mating type information regulation 2 homologue; PPAR, peroxisome proliferator-activated receptor; PGC1a, peroxisome proliferator-activated receptor gamma coactivator-1 alpha; Sglt1, sodium-dependent glucose cotransporter 1; Glut2, glucose transporter 2; Pept1, human peptide transporter 1.

The circadian system has an impact on food digestion, absorption, and metabolism. Furthermore, epithelial cell motility and proliferation in digestive compartments, particularly the colon, show diurnal rhythms. The absorption of glucose and water by the isolated small intestine is higher at night than during the day. The expression of sodium/glucose cotransporter 1 (Sglt1), glucose transporter 2 (Glut2), and glucose transporter 5 (Glut5) has clear circadian oscillations and is regulated by clock genes through E-box activity. Furthermore, PER1 activity regulates Sglt1 independently of the E-box. Following a scheduled feeding experiment, it has been concluded that feeding circumstances directly impact these transporters [9].

Food-derived phenolic compounds can interact with clock genes, which regulate the biological rhythms. In addition, nutrient signaling can affect gut circadian systems. Many important transporters are under circadian regulation, and circadian disruption also leads to abnormal drug absorption [22].

6. Circadian molecular mechanisms modulating the lipid metabolism

It is well known that changing the phase shift by night feeding leads to obesity, the main hypothesis states that the extra energy flow during the rest time is easily converted into lipids accumulation in the adipose tissue. In laboratory practice, it is common to use a high-fat diet to create an animal model of obesity [21, 22]. Although these phenomena are easy to understand, the molecular mechanisms are not fully elucidated.

Abnormalities in the body clock functioning, driven by a high-fat diet, are initiating the weight gain process in mice. This dietary habit also changes the liver clock and the hepatic rhythmicity of lipid metabolism. The influence of meal timing on lipid metabolism is not considered highly significant, while the importance of well-regulated eating habits is recognized [9]. Therefore, the influence of meal timing was examined using genetically unmodified animals. A feeding protocol was developed in which animals ate continuously regardless of time; although restricted feeding (e.g., feeding only during the day) causes day/night reversal in nocturnal rats, they become accustomed to it. It was reported that irregular meals cause abnormalities in the liver circadian clock and increase blood cholesterol levels. It was indicated that differences in meal timing cause abnormalities in cholesterol metabolism, even if the same quantity of food is provided [25]. Hypercholesterolemia was caused by advanced changes in the circadian rhythm and gene expression of CYP7A1, an isoenzyme that limits the rate of bile acid synthesis. Thus, cholesterol metabolism was profoundly altered and bile acid levels excreted decreased. These results indicate that well-regulated dietary habits normalize the liver clock gene and regulate the CYP7A1 rhythm and that blood cholesterol levels are better controlled due to a lower secretion of VLDL (very low-density lipoprotein), namely lowering LDL-cholesterol and raising HDL-cholesterol levels [19].

7. Expanding the perspective of the interplay between chrononutrition and chronotherapy

Time-restricted feeding, a nutritional approach in which food consumption is limited to certain times of the day, allows a daily fasting period of 12 hours or more, thus conferring metabolic resting-frames used by the cellular machinery to initiate and develop complex processes of repair, decreasing the level of accumulated errors due to oxidative stress injuries or aberrant mutations. Understanding the link between time, nutrients, and the benefits of fasting leads to the identification of chrononutritional strategies that mimic fasting and achieve similar changes to those triggered by fasting [3].

Acknowledging the pervasive and constant benefits of time-restricted feeding and fast-mimicking diets, basic science and translational research are willing to transform time-managed fasting-related interventions into complex clinical approaches with a remarkable potential to improve human health.

In the clinical scenery, an accurate identification must be performed for all interactions among drugs, drugs and food supplements, drugs and food and, nevertheless, drugs and genetic and epigenetic factors, all being able to impact the therapeutic outcome considerably. The absorption, distribution, metabolism, and elimination of drugs can be highly influenced by slight variations of the environmental factors and endogenous elements, mainly affecting their bioavailability, efficacy, and metabolism to toxic compounds [26].

The most exposed to this phenomenon are neurologic and cardiovascular patients due to the complexity of their pathology and the particularities of these anatomic and physiologic systems. Restricted scheduled food intake in experimental models determines the occurrence of food anticipatory activity in animals, which is observed approximately 2 hours before the feeding time. The learning process of specific times of feeding is acquired by the internal food-entrainable oscillator mechanism and this food-dependent entraining also encompasses clock gene expression rhythms in major cerebral and peripheral tissues, except the SCN. Several experimental studies demonstrated that by scheduled feeding, the animals depicted Per1, Per2, D-site-binding protein, and cholesterol 7 alpha-hydroxylase mRNA expression rhythms that underwent rapid phase shifting and entrainment at the hepatic level, and a slower rate in kidneys, heart, and pancreas and did not undergo at all scheduled feedings-phase shifting in the SCN [3, 5].

Regarding antihypertensive and neurotropic drugs, when evaluating their therapeutic efficacy, we have to equally consider the time of administration, their precise dose, and eventual matrix effects that can completely change their bioavailability profile. This is mainly due to their direct interdependence with biological rhythms, blood pressure physiologic oscillation during 24 hours, and nevertheless the circadian rate of metabolism at the hepatic level. The nutritional impact can also be displayed by food constituents that have additive or antagonist effects such as phenolic compounds and peptides, in conjunction with blood pressure levels [27].

Chronotherapeutic-chrononutritional studies conducted by Matsunaga et al. are of unique relevance in this field assessing the circadian pattern of hepatotoxicity and mortality rates after acetaminophen administration in mice subjected to ad libitum versus time-restricted feeding patterns, and food-entrained circadian rhythms modulated toxic effects through CYP2E1 and hepatic glutathione activities [28]. Analogous employments of food timing patterns on the chrono-pharmacokinetics were described in the activity of sodium valproate and the nephrotoxicity of gentamicin [29, 30].

Insulin signaling is one of the most important factors for food entrainment, as it directly induces Per2 expression in hepatic tissue and cultured hepatocytes [31]. AMPK, a fundamental cellular energy sensor, is another possible factor for food entrainment, being activated by fasting or low glucose levels. It undergoes phosphorylation and destabilizes CRY1 protein. Nutritional ingredients such as caffeine, an antagonist of the adenosine receptors and an inhibitor of phosphodiesterase, increasing cAMP concentrations, can lead to considerable changes in the circadian system as it was reported to lengthen the circadian clock period in the SCN and peripheral tissues and modulates the behavioral rhythms [32]. Functional nutrition may soon become an increasing topic of relevance in future chronotherapeutic strategies.

In most animal species as in humans, the feeding frames alternate with fasting frames. This specific metabolic picture is predominantly dependent on ketone bodies after a prolonged fasting time, glucose, and glycogen being consumed in the first fragment of the fast. Based on this restricted availability of glucose at the cellular level, many hypotheses were formulated to understand the great impact of fasting in preventing a series of metabolic imbalances and also in accompanying the treatment of chronic diseases, mainly cardiovascular, inflammatory, and oncological ones [16, 25, 33, 34].

8. Chronotherapeutic approaches: recognizing the importance of timing factors in the treatment of neurological diseases and sleep disorders

The suprachiasmatic nucleus (SCN) acquires valuable information from the environment, through input signals, such as the light–dark cycle, and nevertheless from other brain areas. There are several important chemical structures that were studied for their influence on the circadian entrainment mechanism of SCN, namely exogenous melatonin and ramelteon, a powerful selective melatonin MT1/MT2 receptor agonist. When administered during the active phase, they act as non-photic trainers that advance SCN circadian rhythms [35–37].

The field of neurology is an extremely complex one, reuniting a series of chronic neurological disabilities from insomnia, epilepsy, and neuromuscular disorders to degenerative diseases, dementia and brain tumors. As a highly prevalent and neurologic imbalance promoting affection, insomnia is affecting more than 30% of the general population, the sleep disturbances being a consequence of stressful life conditions, shift working, physiological aging, but nevertheless being the first clinical sign of a neurologic impairment stage. Its intimate connection with circadian misalignment is undoubtedly approved, addressing an interesting scientific intersection point in chronopharmacology [38–40].

The barbiturates, namely phenobarbital, were among the first-generation drugs used for insomnia, but due to their high abuse potential and associated risks of overdosing, were replaced by benzodiazepines (lorazepam and triazolam).

This second generation of hypnotics displayed different side effects in the area of cognitive and psychomotor impairment, displaying also phenomena of addiction and tolerance, urging the need for a different therapeutic approach. Benzodiazepine receptor agonists (zolpidem and zopiclone) represented the following generation of hypnotics having non-benzodiazepine chemical structures and exhibiting reduced side effects. After extensive clinical studies, it was concluded that the major drawback of this class was that the sleep induced by these pharmacologic agents is electrophysiologically different from the naturally prompted physiological sleep, as they are severely reducing the rapid eye movement (REM) phases of sleep [7, 41].

Melatonin, the pineal hormone discovered in 1958, is the major endogenous regulator of sleep–wake cycles, its synthesis being initiated by the diminishing photic signal, at sunset, and being stopped upon sensing the first light signal. The pineal synthesis pathway includes the essential amino acid tryptophan and most importantly the neurotransmitter serotonin. It has exceptional characteristics, being highly soluble in lipid medium, simply diffusing through almost every cellular membrane, including the blood–brain barrier, the ultimate frontier for most molecules, even the endogenous ones [13, 36, 42].

The half-life of melatonin is extremely short, around 30 minutes, being immediately enzymatically transformed in a series of metabolites, with particular functions in the oxidative stress array, in order to be finally converted by the liver and then urinary excreted. No other endogenous molecule shares the same unique strong circadian pattern of synthesis in healthy human organisms: the plasma level is detectable immediately after sunset, registering a peak around 3 a.m. and becomes untraceable in the early morning, a period superposable with the light signal initiation [43].

The intimate connection between melatonin and the utmost central nervous system structures is achieved through the specific receptors which are found in high densities in circadian regulatory entities, and in fact, to a certain extent, in most human organs. It is by this mechanism that it is understood the role of melatonin as a master hormone, endogenous synchronizer, and circadian modulator of all biological internal systems. Melatonin mediates the information regarding the dark signals throughout the entire body, conducting chronobiotic and phase shift effects, hypnotic by imprinting sleep–wake robust cycles, but at the same time, having a versatile modulatory ability to adjust the circadian pattern in disruptive circumstances [35].

Considered the initiator and maintainer of sleep in humans, the darkness hormone has proved its sleep-inducing effects only concerning the endogenous molecule, as the exogenous supplementation has resulted in controversial conclusions, mainly due to its short half-life, increased first-pass metabolism, and weak receptors binding. As the major non-photic entrainer of the circadian rhythm, exogenous melatonin is a powerful pharmacologic tool in correcting circadian misalignments in patients [44].

Melatonin was subjected lately to modern pharmaceutical formulations that assure better bioavailability characteristics that were efficient for inducing sleep and increasing the quality and length of sleeping time in elderly patients suffering from chronic insomnia, as it is acknowledged that the endogenous melatonin synthesis declines physiologically by aging, due to pineal gland calcification, decreased sensing of the light signal, or a decline in the SCN function [21].

Chronopharmacological approaches are of remarkable importance for a chronobiotic agent such as melatonin. A series of experimental and clinical studies have assessed the relationship between the administered melatonin dose, the time of administration, and the occurrence of biological effects and their intensity, in a plethora of circadian rhythm-derived sleep disorders, from a jet lag sleep disorder, shift work sleep disorder, delayed sleep phase syndrome, primary insomnia occurring in various psychiatric illness. The conclusions are unanimous stating that the time of the day used for administering exogenous melatonin is indisputably determining the precise effect on the circadian rhythms: delaying them after the morning administration, and on the contrary, advancing the circadian phase and subsequent evening sleep induction, following a late afternoon or night drug administration [40, 45].

Based on the clinical data, but mostly on the evidence suggesting the tremendous therapeutic effect displayed upon selectively binding the melatoninergic receptors, the MT1/MT2 agonist ramelteon was assessed for its role in circadian re-entrainment and for inducing sleep in refractory insomnia. The time of administration is also an important issue, taken into consideration by Watanabe et al. study, which concluded on the efficiency of small doses of 1–2 mg taken at more than 5 hours before bedtime, assumed as an early administration pattern [46]. In the vast majority of the performed trials, ramelteon proved good effects, especially on refractory insomnia cases, often combined with serious disturbances in circadian rhythms of sleep–wake cycles. All scientific data indicate that ramelteon acts not only as a hypnotic agent but also at a molecular level it is mimicking all central effects of melatonin as a circadian entrainer [37, 47, 48].

Numerous other central nervous system-acting drugs are of utmost chronotherapeutic importance due to their particular pharmacologic patterns in connection with their time of administration. For instance, the night-time administration of benzodiazepines down-regulates the expression of Per1 and Per2 genes, having a direct impact on the entrainment process at the level of SCN. Lithium, a classic mood stabilizer prescribed in manic episodes associated with bipolar disorder, inhibits glycogen synthase kinase-3 (GSK-3) in the suprachiasmatic nucleus, increasing the locomotor activity cycle [44, 49]. Anesthetic agents significantly influence circadian rhythms, causing either phase shifts, or diminishing the rhythmic amplitude of clock gene expression. Many central nervous system drugs, when administered together, can affect circadian rhythm via their target receptors and metabolism enzymes [2].

9. Current understandings and future directions in the chronopharmacology management of cardiovascular chronic diseases

Circadian rhythms have tremendous importance in cardiovascular system functions in both healthy people and patients, being one of the first scientifically acknowledged physiopathologic areas interlinked by chronomodulatory mechanisms [25]. Heart rate and blood pressure (BP) rhythms are the most studied periodic functions in the circulatory system, but lately, new evidence suggests blood flow, cardiac output, stroke volume, or peripheral resistance can furthermore trigger relevant circadian changes [50, 51].

Hypertension, the high systemic blood pressure, represents the major risk factor for a series of acute or chronic pathologic events, from myocardial infarction and ischemic events to chronic kidney failure and sudden death. Chronic high blood pressure is a silent condition since it progresses primarily asymptomatically, but has a devastating effect, conducting to cardiovascular disruptive events and end-organ damage, which leads to a reduction in the quality of life or events of life expectancy. The simple association between high blood pressure and cardiovascular diseases, which is largely controlled by our social and nutritional behavior, is also influenced by a circadian pattern. As a consequence, circadian rhythm plays an important role in blood pressure management, which is the reason for chronobiological studies to comprise chronopharmacology and chrononutrition aspects in the same research field [24, 52, 53].

Blood pressure registers circadian variations due to internal and external modulatory factors. Endogenous regulatory performers such as hematological and renal inputs, endocrine signals perceived by the circulatory system, and the autonomic nervous system activity, interact with environmental parameters like temperature and humidity, nutrients intake, vasoactive xenobiotics as alcohol and caffeine, and nevertheless with the physical activity, emotional status, and the sleep pattern. Despite the overabundance of parameters that influence it, circadian variations of blood pressure in normotensive patients are characterized by a decrease in both systolic and diastolic blood pressure during the night, on average by 10–20% compared with diurnal values, defining the normal pressure physiological profile [20, 33].

The magnitude of the decrease in nocturnal BP values was mainly quantified by calculating the dipping index (relative nocturnal decrease in BP) which is defined as the percentage decrease in the value of the average night BP compared with the mean daily BP. This dipping index allowed the classification of the night profile of TA in dipper profile when the index is between 10 and 20%, non-dipper profile when the night decrease of TA is less than 10%, of extreme-dipper type when the decrease of nocturnal BP is over 20% of daytime values and riser profile when BP increases during sleep. The classification has acquired a great utility, especially from the point of view of the prognosis, especially since there are studies that demonstrate the reproducibility of the evolution of nocturnal values of BP, in the same individual, for several months [51].

The circadian rhythm of BP shows maximum values in the early hours of the morning, followed by a subsequent gradual decrease and with minimum nocturnal values, respectively. Explanations of this variability involve the influence of exogenous or environmental factors and endogenous factors.

The endogenous rhythm associated with the body's clock determines amplitudes ranging from 5 to 10 mmHg. Blood pressure increases adjacent to waking time are associated with increases in catecholamine plasma levels: both norepinephrine and epinephrine are at peak levels in the morning and low levels at night [33]. At the moment, there are many drug classes that are used in the management of hypertension, the majority of therapeutic schemes comprising a combination of them, as they have different mechanisms of action and sometimes variable sites of action, being complementary: α- and β-adrenoceptor blocking drugs, calcium channel blockers, diuretics, angiotensin-converting enzyme (ACE) inhibitors, AT1-receptor blockers, etc. (**Figure 3**). Ambulatory blood pressure monitoring of hypertensive patients is of absolute importance as it can reveal abnormal variable patterns predicting super dipping and late-night angina crises [54, 55].

Furthermore, the pharmacokinetic profile, half-life, formulation, duration of pharmacological impact, and hence the dosing interval vary between antihypertensive medication classes and individual compounds.

The primary steps in the processes that regulate blood pressure are reliant on the circadian phase, indicating that β-adrenoceptor antagonists have no effect on, diminish, or even abolish the rhythmic pattern in BP. β-adrenoceptor antagonists, on the other hand, have the property to reduce daytime BP levels and have little effect on night-time values, being not effective in lowering the early morning spike in BP. Heart rate declines caused by β-adrenoceptor antagonists are consistently more evident during the daytime hours [24, 56].

Taking into account numerous studies conducted conventionally, β-adrenoceptor antagonists—β1-selective, nonselective, or with intrinsic sympathomimetic activity do not modify or lessen the rhythmic pattern in blood pressure. Generally, it is estimated that adrenoceptor antagonists lower daytime blood pressure levels while having little or no effect on nocturnal values and are less effective in reducing the early morning escalation in blood pressure levels [56].

The effects of **β-adrenoceptor antagonists** on heart rate are consistently more evident during the daytime. Correspondingly, a fourfold crossover trial using propranolol in healthy adults revealed a more dramatic reduction in heart rate and blood pressure during the day than at night. Furthermore, it is demonstrated that the circadian phase can affect the dose–response relationship. Pindolol, a partial agonist, even raised the heart rate at night, which was a surprising finding [26, 34]. Thus, clinical evidence suggests that β-adrenoceptor-mediated blood pressure regulation is

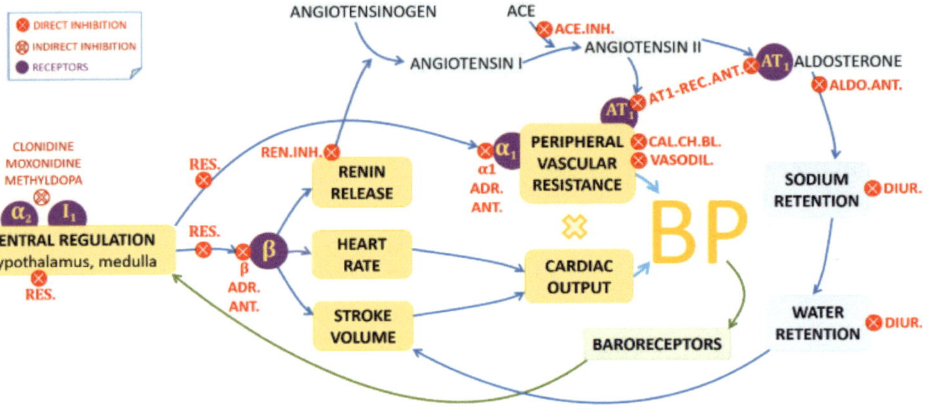

Figure 3.
The biological mechanisms of blood pressure (BP) and the therapeutic approaches that are subjected to circadian modulation. α1 ADR.ANT, α1-adrenoreceptor antagonists; β ADR.ANT, β-adrenoreceptor antagonists; ACE. INH, ACE inhibitors; ALDO.ANT, aldosterone antagonists; AT1-REC.ANT, AT1-receptor antagonists; CAL. CH.BL, calcium channel blockers; DIUR, diuretics; REN.INH, renin inhibitors; RES, reserpine; VASODIL, vasodilators.

more important during the day and less important at night and early in the morning. This is consistent with the circadian regularity in sympathetic tone, as evidenced by plasma noradrenaline and cAMP levels [56].

Calcium channel blockers' effects were also investigated, primarily by the assessment of blood pressure profiles. The administration of non-retard verapamil three times a day did not significantly modify the blood pressure profile in primary hypertensive patients, however, it was less effective at night. A single morning dose of sustained-release verapamil provided good 24-hour blood pressure management, while a sustained-release diltiazem formulation was less effective at night [26, 52]. Different pharmacokinetics of dihydropyridine derivatives appear to decrease blood pressure to wide-ranging degrees during the day and night, the specific drug formulation, and dose interval play a role [27].

In essential hypertensive patients (dippers), amlodipine, sustained-release isradipine and lacidipine, nifedipine gastrointestinal therapeutic system (GITS), and nisoldipine extended-release (ER), as well as in normotensives, immediate-release nifedipine had no impact on the 24-hour blood pressure profile after one dose taken in the morning and one in the evening, whereas, with nitrendipine, there was either no impact or a slight impact after the evening dose. Nifedipine, taken twice a day, reduced blood pressure over a 24-hour period in primary hypertension individuals. Most notably, isradipine only normalized the greatly disturbed blood pressure profile in secondary hypertensives (non-dippers) due to renal failure after the evening, but not the morning dosing. In contrast, amlodipine and nisoldipine ER normalized the disturbed blood pressure profile in non-dippers after both morning and evening dosing [20].

These studies clearly show that calcium channel blockers lower elevated blood pressure in both non-dippers and dippers without distorting the latter's normal blood pressure profile or causing super dipping and can convert non-dipping behavior into dipping behavior, the dose administered in the evening being more appropriate.

Several crossover studies, assessing morning versus evening dose, with **ACE inhibitors** in essential hypertensive patients, revealed that among dipper patients, the evening dosing of benazepril, enalapril, and perindopril resulted in a more dramatic overnight decline than the morning dose, resulting in a super dipping blood pressure profile. Evening quinapril doses had a stronger effect than the morning ones, but the blood pressure pattern was not significantly altered [57].

After either dosing time, ramipril had no visible effect on the 24-hour blood pressure profile. An excessively severe nocturnal reduction in blood pressure, the super dipping pattern, following night-time dosage could be a possible risk factor for the development of ischemic events in individuals with hypertension, given their reduced cardiac reserve.

Other antihypertensive agents have been seldom explored in regard to probable circadian fluctuations. In essential hypertensive patients, once-daily morning doses of the diuretics xipamide and indapamide decreased blood pressure without affecting its 24-hour pattern [58].

Trials conducted on diuretics in salt-sensitive hypertensive patients (dippers and non-dippers), revealed that diuretics did not modify the circadian blood pressure profile in dippers, but did turn non-dippers into dippers [20, 27, 33].

The **α-adrenoceptor antagonists** indoramin and prazosin did not modify the blood pressure profile when administered twice daily. Throughout both day and night, a single night-time dose of the α-adrenoceptor antagonist doxazosin lowered equally systolic and diastolic blood pressure, however, the greatest decrease occurred in the morning hours. Recent research in dippers using doxazosin-gastrointestinal therapeutic system (GITS) as monotherapy demonstrated a slight but considerable reduction in blood pressure over the course of 24 hours without disrupting the normal blood pressure profile. These findings suggest the importance of α-adrenoceptor mediated BP regulation during the early morning hours because α-adrenoceptor blockade lowered peripheral resistance more efficiently during the early morning hours than at other times of the day.

The capacity of the night/day ratio of systolic BP to predict the risk of cardiovascular events is more accurate than BP recorded once, according to studies conducted in recent years. Given that nocturnal blood pressures are most consistently associated with cardiovascular risk, they require closer monitoring for patient safety [53]. The circadian rhythm plays an important role in the regulation of blood pressure, and research conclusions suggest that time is one of the most important factors influencing cardiovascular risk management. As a result, chronopharmacotherapy is required for circadian disorders including hypertension.

It is obvious that circadian rhythms have a significant impact on cardiovascular disorders. Metabolic pathways, signal transduction cascades, transcriptional networks, protein turnover, and other processes are all timed to promote optimal cellular and organ functioning.

The disruption of the circadian governance almost always triggers pathology. In addition, it is more and more obvious that the persistence and augmentation of circadian rhythms might cause and aggravate CVD in vulnerable individuals. For that purpose, it is important to identify all the details regarding the mechanisms implied by the entrainment of the cardiovascular system, using the pattern of cell-to-cell synchronization, and the particular points where the phase misalignment is transformed into pathological events (**Figure 4**).

By gathering as much chronobiologic data as possible for a single patient, it is conceivable to develop therapeutic strategies that specifically target circadian clock components and downstream mediators for the effective treatment and prevention of cardiovascular diseases, to adjust some significant lifestyle behaviors, as the timing of eating, of performing exercise and environmental conditions, as the exposure to light and certain temperatures, that work in synchrony with circadian governance of cellular processes [2, 53].

Food and Drug Administration recommends that the vast majority of long-acting formulations are preferable to be administered in the evening, at bedtime. Exceptions are represented beta-blockers that are not specifically designed for chronotherapy or do not have a similar 24-hour effect from the point of view of heart rate reduction making sense to administer them in the morning when the sympathetic drive is predominant, diuretics have the same effect over 24 hours, but the discomfort from frequent night-time awakenings should be avoided. In addition, quinapril at doses of 30–40 mg may register an excessive effect and doxazosin may provide an excessive decrease when administered in the evening. There are indications for the administration of antihypertensive drugs early in the morning, for the dipper patients, and for the non-dippers, it is necessary to add an evening dose or to switch to a single evening dose, in order to acquire a reduction in the blood pressure levels and a normalization of the 24-hour profile that is profoundly disturbed in these patients [33]. As cardiovascular drugs have vulnerable pharmacokinetic profiles that can be easily influenced by the circadian phase, it is of great importance to consider the half-life of the active components and the exact parameters of the employed pharmaceutical formulation, in order to emit a sustainable conclusion regarding the best dosage time within 24 hours [15, 26].

All the data converge to show that for the "non-dipper" type of patients and for "extreme dippers" the cardiovascular prognosis is rather poor. However, the only way

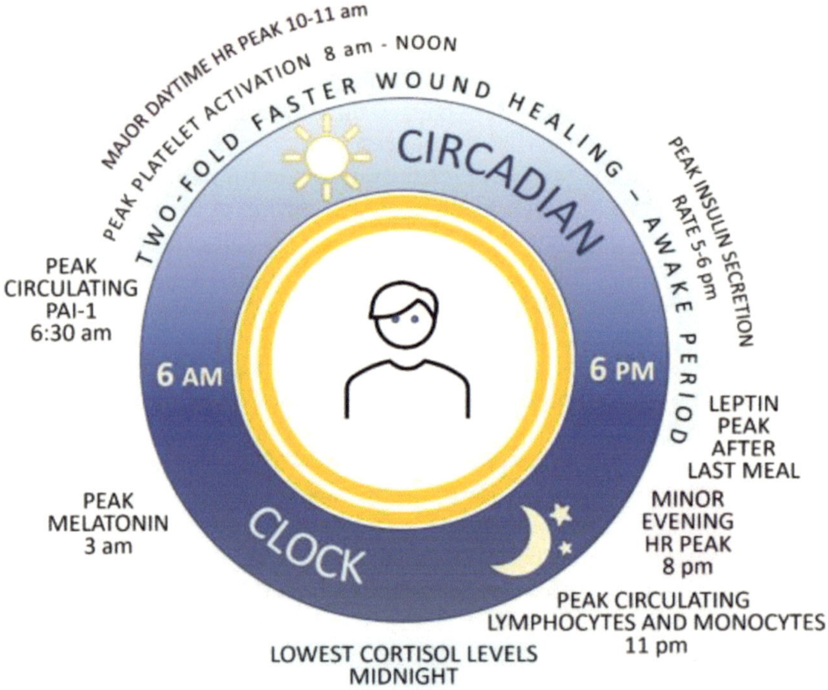

Figure 4.
The circadian imprint upon physiologic endocrine and metabolic parameters often reflected in the pathologic scenery of the human organism.

to positively influence these unfavorable types of circadian variability of BP values can only be achieved by chronotherapy. It offers a way to personalize the treatment of hypertension considering each patient's distinct circadian profile, providing a better perspective on optimal blood pressure control and decreasing the associated cardiovascular risk.

It must be acknowledged that the development of antihypertensive therapy provides an effective arsenal in the sense of chronotherapy: the evolution from short-acting drugs to drugs that release the active substance slowly, allowing them to be administered in such a way as to achieve maximum effectiveness, both in terms of ideal day-time blood pressure values and by inducing a dipper-like profile of circadian variability during night-time.

10. Conclusions

In essence, food intake performs a strong entrainable role in regard to human circadian rhythms. In addition, this entrainment modulates the timing of food digestion and metabolism by controlling clock-regulated output genes in the peripheral tissues, having tremendous effects on drug bioavailability and metabolism.

Understanding the mechanisms of food entrainment in the circadian system and the complexity of nutritional signals will contribute to chrono-nutritional therapy guidelines concerning the joint functionality of food and nutrition.

In the next period of time, further research will completely elucidate the interactions along the remarkable time–nutrition–drug axis within the framework of chrononutrition modulated chronopharmacotherapy, becoming an essential tool for chronic pathologies management, by individualizing the chrono-pharmaco-therapeutic approach for every single patient.

Author details

Alina Crenguța Nicolae[1], Ion-Bogdan Dumitrescu[1*], Camelia Cristina Diaconu[1,2], Mirela Elena Ritivoiu[1,3], Carmen Adella Sirbu[1,4] and Cristina Manuela Drăgoi[1]

1 "Carol Davila" University of Medicine and Pharmacy, Bucharest, Romania

2 Clinical Emergency Hospital, Internal Medicine Clinic, Bucharest, Romania

3 "Alessandrescu-Rusescu" Mother and Child's Health National Institute, Bucharest, Romania

4 Central Military Emergency University Hospital, Neurology Clinic, Bucharest, Romania

*Address all correspondence to: ion.dumitrescu@umfcd.ro

References

[1] Lemmer B, Portaluppi F. Chronopharmacology of cardiovascular diseases. In: Redfern P, Lemmer B, editors. Handbook of Experimental Pharmacology, Physiology and pharmacology of biological rhythms. Vol. 125. New York: Springer; 1997. pp. 251-297

[2] Dobrek L. Chronopharmacology in therapeutic drug monitoring—dependencies between the rhythmics of pharmacokinetic processes and drug concentration in blood. Pharmaceutics. 2021;**13**(11):1915. DOI: 10.3390/pharmaceutics13111915

[3] Tahara Y, Shibata S. Chrono-biology, chrono-pharmacology, and chrono-nutrition. Journal of Pharmacological Sciences. 2014;**124**(3):320-335. DOI: 10.1254/jphs.13r06cr

[4] Dragoi CM, Morosan E, Dumitrescu IB, Nicolae AC, Arsene AL, Draganescu D, et al. Insights into chrononutrition: The innermost interplay amongst nutrition, metabolism and the circadian clock, in the context of epigenetic reprogramming. Farmácia. 2019;**67**:557-571

[5] Gloston GF, Yoo SH, Chen ZJ. Clock-enhancing small molecules and potential applications in chronic diseases and aging. Frontiers in Neurology. 2017;**8**:100

[6] Musiek ES, Holtzman DM. Mechanisms linking circadian clocks, sleep, and neurodegeneration. Science. 2016;**354**:1004-1008

[7] Fifel K, Videnovic A. Circadian alterations in patients with neurodegenerative diseases: Neuropathological basis of underlying network mechanisms. Neurobiology of Disease. 2020;**144**:105029

[8] Sion B, Bégou M. Can chronopharmacology improve the therapeutic management of neurological diseases? Fundamental & Clinical Pharmacology. 2021;**35**:564-581

[9] Ahmed N, Farooq J, Siddiqi HS, Meo SA, Kulsoom B, Laghari AH, et al. Impact of intermittent fasting on lipid profile—A quasi-randomized clinical trial. Frontiers in Nutrition. 2021;**7**:596787

[10] Scheving LA. Biological clocks and the digestive system. Gastroenterology. 2000;**119**:536-549

[11] Voiculescu SE, Le Duc D, Roşca AE, Zeca V, Chiţimuş DM, Arsene AL, et al. Behavioral and molecular effects of prenatal continuous light exposure in the adult rat. Brain Research. 2016;**1650**:51-59

[12] Zisapel N. New perspectives on the role of melatonin in human sleep, circadian rhythms and their regulation. British Journal of Pharmacology. 2018;**175**:3190-3199

[13] Pevet P, Challet E. Melatonin: Both master clock output and internal time-giver in the circadian clocks network. Journal of Physiology, Paris. 2011;**105**:170-182

[14] Bron R, Furness JB. Rhythm of digestion: Keeping time in the gastrointestinal tract. Clinical and Experimental Pharmacology & Physiology. 2009;**36**:1041-1048

[15] Challet E. The circadian regulation of food intake. Nature Reviews. Endocrinology. 2019;**15**:393-405

[16] Polidarová L, Soták M, Sládek M, Pacha J, Sumová A. Temporal gradient in

the clock gene and cell-cycle checkpoint kinase Wee1 expression along the gut. Chronobiology International. 2009;**26**:607-620

[17] Hoogerwerf WA, Sinha M, Conesa A, Luxon BA, Shahinian VB, Cornélissen G. Transcriptional profiling of mRNA expression in the mouse distal colon. Gastroenterology. 2008;**135**:2019-2029

[18] McClung CA. Mind your rhythms: An important role for circadian genes in neuroprotection. The Journal of Clinical Investigation. 2013;**123**:4994-4996

[19] Huxley RR, Barzi F, Lam TH, Czernichow S, Fang X, Welborn T, et al. Isolated low levels of high-density lipoprotein cholesterol are associated with an increased risk of coronary heart disease. Circulation. 2011;**124**(19):2056-2064

[20] Hermida RC, Hermida-Ayala RG, Mojón A, Smolensky MH, Fernández JR. Systematic review and quality evaluation of published human ingestion-time trials of blood pressure-lowering medications and their combinations. Chronobiology International. 2021;**38**(10):1460-1476

[21] Dragoi C.M, Nicolae A.C., Grigore C., Dinu-Pîrvu C.E., Arsene A.L., Characteristics of glucose homeostasis and lipidic profile in a hamster metabolic syndrome model, after the co-administration of melatonin and irbesartan in a multiparticulate pharmaceutical formulation. The Second International Conference on Interdisciplinary Management of Diabetes Mellitus and its Complications, INTERDIAB 2016; 2016.Niculescu, Bucharest: Mar 3-5, pp. 221-229. Diabetes Mellitus as Cardiovascular Disease.

[22] Kohsake A, Laposky AD, Ramsey KM, Estrada C, Joshu C, Kobayashi Y, et al. High-fat diet disrupts behavioral and molecular circadian rhythms in mice. Cell Metabolism. 2007;**6**(5):414-421

[23] Hoogerwerf WA, Hellmich HL, Cornélissen G, Halberg F, Shahinian VB, Bostwick J, et al. Clock gene expression in the murine gastrointestinal tract: Endogenous rhythmicity and effects of a feeding regimen. Gastroenterology. 2007;**133**(4):1250-1260

[24] Chen Z, Yoo SH, Takahashi JS. Development and therapeutic potential of small-molecule modulators of circadian systems. Annual Review of Pharmacology and Toxicology. 2018;**58**:231-252

[25] Ahmet I, Wan R, Mattson MP, Lakatta EG, Talan M. Cardioprotection by intermittent fasting in rats. Circulation. 2005;**112**(20):3115-3121

[26] Bicker J, Alves G, Falcao A, Fortuna A. Timing in drug absoption and disposition: The past, present, and future of chronopharmacokinetics. British Journal of Pharmacology. 2020;**177**(10):2215-2239

[27] Smolensky MH, Siegel RA, Haus E. Biological rhythms, drug delivery, and chronotherapeutics. In: Siepmann J, Siegel R, Rathbone M, editors. Fundamentals and Applications of Controlled Release Drug Delivery. Advances in Delivery Science and Technology. Boston, MA: Springer; 2012. pp. 359-443

[28] Matsunaga N, Nakamura N, Yoneda N, Qin T, Terazono H, To H. Influence of feeding schedule on 24-h rhythm of hepatotoxicity induced by acetaminophen in mice. The Journal of Pharmacology and Experimental Therapeutics. 2004;**311**:594-600

[29] Beauchamp D, Guimont C, Grenier L, LeBrun M, Tardif D,

Gourde P, et al. Timerestricted feeding schedules modify temporal variation of gentamicin experimental nephrotoxicity. Antimicrobial Agents and Chemotherapy. 1997;**41**:1468-1474

[30] Ohdo S, Ogawa N, Nakano S, Higuchi S. Influence of feeding schedule on the chronopharmacological aspects of sodium valproate in mice. The Journal of Pharmacology and Experimental Therapeutics. 1996;**278**:74-81

[31] Yamajuku D, Inagaki T, Haruma T, Okubo S, Kataoka Y, Kobayashi S. Real-time monitoring in three-dimensional hepatocytes reveals that insulin acts as a synchronizer for liver clock. Scientific Reports. 2012;**2**:439

[32] Oike H, Kobori M, Suzuki T, Ishida N. Caffeine lengthens circadian rhythms in mice. Biochemical and Biophysical Research Communications. 2011;**410**:654-658

[33] Bowles NP, Thosar SS, Herzig MX, Shea SA. Chronotherapy for hypertension. Current Hypertension Reports. 2018;**20**:97. DOI: 10.1007/s11906-018-0897-4

[34] Wallach T, Kramer A. Chemical chronobiology: Toward drugs manipulating time. FEBS Letters. 2015;**589**:1530-1538

[35] Garfinkel D, Laudon M, Zisapel N. Improvement of sleep quality by controlled-release melatonin in benzodiazepine-treated elderly insomniacs. Archives of Gerontology and Geriatrics. 1997;**24**:223-231

[36] Medeiros CA, Carvalhedo de Bruin PF, Lopes LA, Magalhaes MC, de Lourdes SM, de Bruin VM. Effect of exogenous melatonin on sleep and motor dysfunction in Parkinson's disease. A randomized, double blind, placebo-controlled study. Journal of Neurology. 2007;**254**:459-464

[37] Miyamoto M. Pharmacology of ramelteon, a selective MT1/MT2 receptor agonist: A novel therapeutic drug for sleep disorders. CNS Neuroscience & Therapeutics. 2009;**15**:32-51

[38] Logan RW, McClung CA. Rhythms of life: Circadian disruption and brain disorders across the lifespan. Nature Reviews. Neuroscience. 2019;**20**:49-65

[39] Radu G, Bordejevic AD, Buda V, Tomescu MC, Dragan I, Dehelean L, et al. Cardiovascular risk factors for different types of psychiatric pathologies: A correlative study. Farmácia. 2020;**68**:5

[40] Reiter R, Sharma R, Rosales-Corral S, Manucha W, de Almeida Chuffa LG, de Campos Zuccari DAP. Melatonin and pathological cell interactions: Mitochondrial glucose processing in cancer cells. International Journal of Molecular Sciences. 2021;**22**(22):12494. DOI: 10.3390/ijms222212494

[41] Yokota SI, Horikawa K, Akiyama M, Moriya T, Ebihara S, Komuro G. Inhibitory action of brotizolam on circadian and light-induced per1 and per2 expression in the hamster suprachiasmatic nucleus. British Journal of Pharmacology. 2000;**131**:1739-1747

[42] Estrada-Reyes R, Valdés-Tovar M, Arrieta-Baez D, Dorantes-Barrón AM, Quero-Chávez D, Solís-Chagoyán H, et al. The timing of melatonin administration is crucial for its antidepressant-like effect in mice. International Journal of Molecular Sciences. 2018;**19**(8):2278

[43] Dowling GA, Mastick J, Colling E, Carter JH, Singer CM, Aminoff MJ.

Melatonin for sleep disturbances in Parkinson's disease. Sleep Medicine. 2005;**6**:459-466

[44] Haimov I, Lavie P, Laudon M, Herer P, Vigder C, Zisapel N. Melatonin replacement therapy of elderly insomiacs. Sleep. 1995;**18**:598-603

[45] Riemersma-van der Lek RF, Swaab DF, Twisk J, Hol EM, Hoogendijk WJ, Van Someren EJ. Effect of bright light and melatonin on cognitive and noncognitive function in elderly residents of group care facilities: A randomized controlled trial. Journal of the American Medical Association. 2008;**299**:2642-2655

[46] Watanabe A, Hirose M, Kitajima T. A retrospective study of the efficacy of ramelteon for insomnia: Relevance of dose and timing of administration. Sleep and Biological Rhythms. 2018;**16**:69-75

[47] Pandi-Perumal SR, Srinivasan V, Poeggeler B, Hardeland R, Cardinali DP. Drug insight: The use of melatonergic agonists for the treatment of insomnia-focus on ramelteon. Nature Clinical Practice. Neurology. 2007;**3**:221-228

[48] Rawashdeh O, Hudson RL, Stepien I, Dubocovich ML. Circadian periods of sensitivity for ramelteon on the onset of running-wheel activity and the peak of suprachiasmatic nucleus neuronal firing rhythms in C3H/HeN mice. Chronobiology International. 2011;**28**:31-38

[49] Li J, Lu WQ, Beesley S, Loudon AS, Meng QJ. Lithium impacts on the amplitude and period of the molecular circadian clockwork. PLoS One. 2012;**7**(3):e33292

[50] Barbu CG, Arsene AL, Florea S, Albu A, Sirbu A, Martin S, et al. Cardiovascular risk assessment in osteoporotic patients using osteoprotegerin as a reliable predictive biochemical marker. Molecular Medicine Reports. 2017 Nov;**16**(5):6059-6067. DOI: 10.3892/mmr.2017.7376.

[51] Boggia J, Li Y, Thijs L, Hansen TW, Kikuya M, BjörklundBodegård K, et al. Prognostic accuracy of day versus night ambulatory blood pressure: A cohort study. Lancet. 2007;**370**(9594):1219-1229

[52] Diaconu CC, Cozma MA, Dobrică EC, Gheorghe G, Jichitu A, Ionescu VA, et al. Polypharmacy in the management of arterial hypertension—Friend or foe? Medicina. 2021;**57**:1288

[53] Diaconu CC, Drăgoi CM, Bratu OG, Neagu TP, Pantea Stoian A, Cobelschi PC, et al. New approaches and perspectives for the pharmacological treatment of arterial hypertension. Farmácia. 2018;**66**(3):408-415

[54] Booth JN, Muntner P, Abdalla M, Diaz KM, Viera AJ, Reynolds K, et al. Differences in night-time and daytime ambulatory blood pressure when diurnal periods are defined by self-report, fixed times, and actigraphy: Improving the detection of hypertension study. Journal of Hypertension. 2016;**34**(2):235-243

[55] Portaluppi F, Tiseo R, Smolensky MH, Hermida RC, Ayala DE, Fabbian F. Circadian rhythms and cardiovascular health. Sleep Medicine Reviews. 2012;**16**:151-166

[56] Fares A. Night-time exogenous melatonin administration may be a beneficial treatment for sleeping disorders in beta blocker patients. Journal of Cardiovascular Disease Research. 2011;**2**(3):153-155

[57] Rodan (Pura) G, Nastasă C, Marc G, Simu H, Oniga O. Evaluation of the efficacy of the association of urapidil

with other antihypertensive drugs in the prehospital emergency in Romania. Farmácia. 2019;**67**:5

[58] Gavrilescu CM, Cernomaz A, Crisan-Dabija R, Popescu RI, Barbu RM, Boisteanu D, et al. A pharmacoepidemiological study on the side effects of indapamide. Farmácia. 2020;**68**(5):932-939